SpringerBriefs in Applied Sciences and Technology

SpringerBriefs present concise summaries of cutting-edge research and practical applications across a wide spectrum of fields. Featuring compact volumes of 50 to 125 pages, the series covers a range of content from professional to academic.

Typical publications can be:

- A timely report of state-of-the art methods
- An introduction to or a manual for the application of mathematical or computer techniques
- A bridge between new research results, as published in journal articles
- A snapshot of a hot or emerging topic
- An in-depth case study
- A presentation of core concepts that students must understand in order to make independent contributions

SpringerBriefs are characterized by fast, global electronic dissemination, standard publishing contracts, standardized manuscript preparation and formatting guidelines, and expedited production schedules.

On the one hand, **SpringerBriefs in Applied Sciences and Technology** are devoted to the publication of fundamentals and applications within the different classical engineering disciplines as well as in interdisciplinary fields that recently emerged between these areas. On the other hand, as the boundary separating fundamental research and applied technology is more and more dissolving, this series is particularly open to trans-disciplinary topics between fundamental science and engineering.

Indexed by EI-Compendex, SCOPUS and Springerlink.

More information about this series at http://www.springer.com/series/8884

Asad Ahmed · Osman Hasan · Falah Awwad ·
Nabil Bastaki

Formal Analysis of Future Energy Systems Using Interactive Theorem Proving

 Springer

Asad Ahmed [ORCID]
School of Electrical Engineering
and Computer Science (SEECS)
National University of Sciences
and Technology (NUST)
Islamabad, Pakistan

Osman Hasan [ORCID]
School of Electrical Engineering
and Computer Science (SEECS)
National University of Sciences
and Technology (NUST)
Islamabad, Pakistan

Falah Awwad [ORCID]
Department of Electrical Engineering
United Arab Emirates University
Al-Ain, United Arab Emirates

Nabil Bastaki
Department of Electrical Engineering
United Arab Emirates University
Al-Ain, United Arab Emirates

ISSN 2191-530X   ISSN 2191-5318 (electronic)
SpringerBriefs in Applied Sciences and Technology
ISBN 978-3-030-78408-9   ISBN 978-3-030-78409-6 (eBook)
https://doi.org/10.1007/978-3-030-78409-6

This Springer imprint is published by the registered company Springer Nature Switzerland AG
The registered company address is: Gewerbestrasse 11, 6330 Cham, Switzerland

To our families

Preface

Smart grids are large and complex networks which mainly employ information and communications technologies (ICT) to ensure cost-effective, efficient, secure and reduced carbon emission energy utilization. Traditionally, paper-and-pencil and simulation methods are used to analyze various aspects of smart grid functionalities. However, these techniques suffer from severe, inherent, limitations in terms of scalability and accuracy. This book presents a theorem proving based logical framework for conducting an accurate analysis of safety- or mission-critical aspects of smart grids. The analysis using this logical framework is highly reliable due to the sound and complete nature of the theorem-proving technique.

The book presents the higher-order-logic formalizations of stability, microeconomics models of cost and utility function and asymptotic theory. These formalizations provide higher-order logic models of the aforementioned theories and their formally verified characteristics within the sound core of the HOL Light theorem prover. These formalizations are then employed to formally verify power converter design, cost and utility models and asymptotic bounds of online scheduling algorithms in smart grids.

The book starts with a brief introduction (Chap. 1) to smart grids in the perspective of energy processing, microeconomics modeling and algorithm design requirements. An overview of the HOL Light theorem prover and its library formalization related to the proposed formalizations is presented (Chap. 2). The formalizations and smart grid applications are presented in the rest of the book (Chaps. 3–5). Finally, Chap. 6 concludes the book.

The target audience of this book are engineers and scientists working in the domains of system analysis and formal methods. These system analysis experts would be able to learn about the potential applications of formal methods, especially, in the safety- and mission-critical applications of smart grids. On the other hand, the online availability of the mechanized proofs provides an opportunity to practitioners to apply these generic formalizations to many other safety- and mission-critical engineering applications, which use these mathematical notions. The whole idea of using theorem proving for smart grids has a great potential to ensure the safe and secure implementation of these state-of-the-art future energy systems.

Islamabad, Pakistan   Asad Ahmed
Islamabad, Pakistan   Osman Hasan
Al-Ain, UAE   Falah Awwad
Al-Ain, UAE   Nabil Bastaki
April 2021

Contents

Acronyms

ABC	Artificial Bee Colony Algorithm
ATP	Automated Theorem Proving
AVR	Average Rate
DER	Distributed Energy Resources
DR	Demand Response
ELF	Expected Load Flattening
EV	Electric Vehicle
FSM	Finite State Machines
ICT	Information and Communications Technology
ITP	Interactive Theorem Proving
LCF	Logic Computable Functions
MPC	Model Predictive Control
OA	Optimal Available
ORCHARD	Online cooRdinated CHARging Decision algorithm
PEV	Plug-in Electric Vehicle
SG	Smart Grid
TF	Transfer Function

Chapter 1
Introduction

Smart grids are future energy networks which utilize state-of-the-art analysis techniques to ensure safe and secure grid operations. These techniques employ a variety of mathematical concepts, such as stability, microeconomics and algorithm design, to solve smart grid problems to achieve the objectives of cost reduction, energy efficiency, quality of service, power mitigation and environment-friendly energy generation. Traditionally, paper-and-pencil and computer-based tools are used to analyze and verify smart grid problems. However, these techniques cannot accurately model and exhaustively verify complex behaviors of systems involving physical and continuous aspects. Smart grids have several components that exhibit continuous behaviors, such as the behavior of underlying electronic components and the impact of weather on renewable sources, and thus they cannot be analyzed completely by the traditional analysis methods. Given their safety- and mission-critical nature, this is a severe limitation as missing a corner case during the smart grid analysis may result in a huge financial loss or even loss of human lives in the extreme cases. To overcome the issues pertaining to the traditional techniques, in this chapter, we present a theorem proving based methodology to formally analyze and specify safety- and mission-critical aspects of smart girds.

1.1 Introduction

Generation and utilization techniques of energy are an important marker of progress of human civilizations [1]. Industrial and information era are two recent examples in human history that are mainly driven by the explorations in the energy sector. With the advent of information and communication era, energy demands have witnessed an enormous increase in the last few decades due to an emerging energy-consuming

© The Author(s), under exclusive license to Springer Nature Switzerland AG 2022
A. Ahmed et al., *Formal Analysis of Future Energy Systems Using Interactive Theorem Proving*, SpringerBriefs in Applied Sciences and Technology,
https://doi.org/10.1007/978-3-030-78409-6_1

human culture [2]. Energy consumptions in household chores, portable devices, data centers [3], electric transportation [4] and industry [8] require solutions that can meet the energy demand efficiently and in an environment-friendly way. Smart grid (SG) [10] has emerged as a viable solution to this challenge as it promises a secure, cost-efficient, environment-friendly and energy-efficient solution to complex and ever-increasing energy demands.

Smart grids are composed of complex networks with intelligent nodes to produce, consume and share the energy efficiently by leveraging upon the advances in the fields of communication, electronics and computation. There has been an enormous increase in the usage of smart grid technology over the world in the last decade or so [11]. Energy harvesting from unconventional sources, such as wind turbines and solar panels, and processing of this energy is one of the key challenges in smart grids due to the intermittent nature of the produced energy [12]. To achieve a steady flow from these sources, power converters are designed and employed in smart grids to alleviate this problem. This objective is usually achieved by designing efficient current and voltage controllers for these power converters so that a smooth supply of power can be ensured. Another key feature of smart grids is to generate, transmit and distribute energy in a cost-effective manner. However, this is not achieved by a single grid action or operation. In this regard, microeconomics models and concepts can be leveraged upon to solve the problems in the energy sector [14, 15], particularly in the smart grids [16], and allow to regulate and trade electricity within the smart grid network [17–21, 82]. In the electricity market, the cost function is used to model the cost of generating, transmitting and distributing the electricity to end-consumers [22]. On the other hand, utility modeling plays a key role in the demand response (DR) programs [23], which are used to shape/reshape the demand of the consumers for safe and secure smart grid operations. Lastly, many smart grid features, such as real-time pricing [24] and online schedules for charging/discharging of plug-in electric vehicles (PEVs) [25, 35, 36, 42], can only be materialized with the help of computationally efficient algorithms. In this context, asymptotic analysis of algorithms plays a vital role in the design of low computational algorithms.

Smart grid deployment and operations are safety- and mission-critical tasks. Power outages and interruptions lead to huge financial losses for the economies all around the globe. In the US, power outages and interruptions cost citizens at least $150 billion each year [37]. In China, electricity shortage in 2004 caused an estimated 0.64% decrease in China's GDP growth [38]. In 2006, 20 countries in Western and Eastern Europe and North Africa suffered approximately $100 million due to just a 2-hour power outage [39]. Moreover, smart grids for restoration of critical loads [40], such as hospitals and street lighting, demand highly reliable grid operations to avoid catastrophic events. Therefore, in this chapter, we have proposed formal methods to formally analyze and verify safety- and mission-critical aspects of smart grids.

Formal methods [41] ensure a highly reliable and accurate analysis and verification of systems due to their computer-based mathematical nature. Formal methods, although comparatively new, have been widely adopted for the analysis of several safety- and mission-critical applications [43–46]. The main idea behind formal methods-based analysis is to develop a mathematical model of the given system and then use mathematical reasoning to deduce that this model exhibits the desired properties in a computer-based sound tool. The first step of formally modeling, usually referred to as formalization, the given system requires the availability of formalized foundational theories of mathematics in the considered logic. In this book, we have formalized several foundational theories for smart grid analysis, namely stability theory (Chap. 3), microeconomics models of cost and utility (Chap. 4) and asymptotic theory (Chap. 5), to formally verify safety- and mission-critical aspects of smart grids.

In the rest of this chapter, we give a brief overview of the smart grids (Sect. 1.2), stability theory (Sect. 1.3), microeconomics modeling (1.4), asymptotic theory (Sect. 1.5), traditional analysis techniques, (Sect. 1.6) and formal methods (Sect. 1.7). Then, in Sect. 1.8, we present a theorem proving, which is a widely used formal methods technique-based methodology to formally analyze and verify smart grids.

1.2 Future Energy Systems: Smart Grids

Bidirectional flow of energy and information among network nodes distinguishes smart grids from the conventional power grids. Figure 1.1 describes the main functionalities of smart grids that are interlinked through a bidirectional information technology and communication layer. This layer is the primary enabler of information and energy processing and a two-way flow in smart grids. In this regard, information and communication technologies furnish the network with energy or data processing and connectivity features. This allows to embed smart infrastructure, management and protection into the future grids [47]. Smart infrastructure includes energy subsystems, such as plug-in electric vehicles [63], information subsystems, such as smart metering and measuring equipment [49], and communication subsystems, such as wired and wireless [49]. Availability of data, in the network, and computational capabilities allows to implement smart management and protection strategies. This leads to an overall complex network, which requires state-of-the-art hardware, software and technology support to achieve the desired functionalities. Consequently, the analysis and verification of these complex systems is also very challenging.

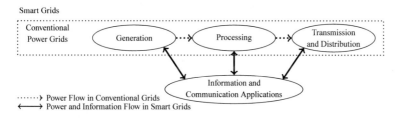

Fig. 1.1 Smart grids functions

Table 1.1 Smart grid functions and analysis methods

Smart grid functions	Analysis			Objectives
	Microeconomics	Stability	Algorithm design	
Integration of DER[a,b,c] [12, 13]	×	✓	✓	Cost, Reliability, Efficiency, Security, Reduced CO_2 emission, Quality of serivces
DR programs[a,b,c] [27, 28, 36]	✓	✓	✓	
Electricity market[a,b] [31–33]	✓	✓	×	

[a]Generation, [b]Transmission and Distribution, [c]Data and energy Processing.

Due to high connectivity, smart grids can achieve their objectives of cost reduction, efficiency, reliability, security and environment friendliness by using smart control and management over generation, transmission and distribution and energy and data processing in the network. At the energy generation level, integration of distributed energy resources (DER), such as alternative or renewable energy sources, electric vehicles and storage systems, allows to mitigate the power outages at the reduced cost and in an environment-friendly manner [5]. The overall cost of the energy is also reduced at the distribution and consumer level by using DR programs [23]. On the other hand, due to the availability of consumers' demand information, smart grids also greatly affect the electricity market economy [14]. Smart grids utilize a broad mix of analysis methods at different levels to achieve these multi-faceted objectives. Table 1.1 shows the utilization of microeconomics, stability and algorithm design analysis in smart grid functions. Many key smart grid functions use these methods as a key ingredient to conduct the analysis and design of the given problems in the network.

This book focuses on the formal verification of smart grid functions that use microeconomics, stability and algorithm design in the design of problems arising in the network. We utilize formal methods to ensure the correctness of the analysis of crucial aspects of smart grid networks using the aforementioned analysis methods. In this regard, we present a generic formalization of a mathematical analysis framework for the stability, microeconomics models and algorithm design, which are defined in the next three sections one-by-one.

1.3   Stability

In smart grids, stability analysis [50] is a mandatory requirement to design power electronic circuits that ensures the smooth flow or harvesting from the alternative or renewable energy sources [6, 12]. Stability analysis is based upon the finding of roots or eigenvalues of the underlying model of the energy system, which is an

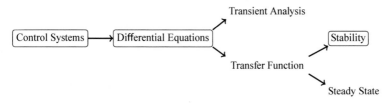

Fig. 1.2 Stability analysis of control systems

essential step for determining the effects of load variances [7] and designing cost and utility models in smart grids [23, 31, 31, 33]. In this section, we provide the basic mathematical foundations of stability on which we develop the formalization of stability in Chap. 3. Moreover, the concepts described here will also be employed, in Chap. 4, for the formal verification of cost and utility modeling in smart grids problems.

Control system theory is used to design control systems, which are usually characterized as systems that are composed of several subsystems connected in an open- or closed-loop configuration and aimed at changing the output of the system as desired. Control system theory is based on the frequency-domain representation of the systems that are obtained using the transform methods, such as Laplace and Fourier, to transform the time-domain description of systems. This allows using several mathematical properties of the frequency-domain representation to ease the analysis of the systems. For example, in frequency domain, the long-term behavior of the system can be expressed at $s = 0$ which corresponds to $t = \infty$ in time domain and convolution property of the time domain is reduced to algebraic multiplication in frequency domain, and, hence the mathematical analysis of systems becomes considerably simple. Moreover, frequency-domain representation allows to specify the system behavior as a transfer function (TF), as shown in Fig. 1.2, which relates the output and input of the system as an algebraic fraction,

$$TF(s) = \frac{O(s)}{I(s)} = \frac{a_m s^m + a_{m-1} s^{m-1} + \cdots + a_0}{b_n s^n + b_{n-1} s^{n-1} + \cdots + b_0}. \qquad (1.1)$$

The numerator and denominator of (1.1) are polynomials, in complex variable s, representing the output and input of the system, respectively, whereas the coefficients a_i and b_i of the polynomials represent system parameters. The roots of these polynomials facilitate the design and analysis of the control systems. The roots of the numerator polynomial are termed as zeros of the system, whereas roots of the denominator polynomial are called poles of the system. Zeros are the values of the complex variable for which the system response is zero and poles are the values of the complex variable for which the system behavior is undefined.

Stability of a system is analyzed using the denominator polynomial of (1.1), also referred to as a characteristic equation, i.e.,

$$b_n s^n + b_{n-1} s^{n-1} + \cdots + b_0 = 0. \qquad (1.2)$$

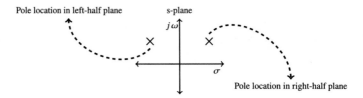

Fig. 1.3 Location of stable roots in s-plane

The roots or eigenvalues of (1.2) characterize the stability of the system. The roots of the characteristic equation are classified as stable, unstable and marginally stable. A stable root lies in the left half s-plane, unstable root lies in the right half of the s-plane and marginally stable root lies on the imaginary axis of the s-plane, as shown in Fig. 1.3.

The mathematical background of stability analysis in this section is central to the proposed formal method-based methodology in this book. Chapters 3 and 4 will cover the detailed account of the formal stability and microeconomics modeling using the theorem-proving technique, mainly using the background presented in this section.

1.4 Cost and Utility Modeling in Microeconomics

In smart grids, DR programs use microeconomics notions, i.e., cost and utility modeling, to design incentives for consumers to change their energy consumption behaviors [27, 29]. Cost function modeling is also used to solve the power flow [30], electricity dispatch [31] problems and spot power prices in the electricity market [34]. In this section, we provide a brief introduction of microeconomics-based cost and utility modeling and related mathematical foundations to help understanding the proposed formalization in Chap. 4.

1.4.1 Microeconomics Modeling

Microeconomics is the study of the economy markets using the models of the economy agents, which are classified as individuals and firms. An individual is termed as a consumer, and a group of well-defined individuals is known as a firm. Furthermore, consumable products are defined as goods, and delivery of these goods is termed as services. In an economy, the production and the consumption of the goods are two main activities which shape an economy market. Microeconomics modeling is primarily aimed at capturing the behaviors of the individuals and firms to consume or offer services so that the overall behavior of the economy can be assessed, such as market prices and profits, based on the decisions made by the individuals and firms [53]. Behavioral economics [54] and rational choice theory [55] are two main approaches for modeling the decisions and choices, i.e., goods or services, of consumers and firms. Behavioral economics incorporates many factors, such as knowledge or information and cognitive capabilities, that affect the decision-making of a consumer or a firm. On the other hand, rational choice theory is an axiomatic approach which assumes that the decision makers, i.e., an individual or group of individual, are always able to prefer and rank the given set of choices. Similarly, the non-decreasing property of the utility modeling function ensures that the choices and preferences are always intended to maximize the happiness or satisfaction of the individual. Moreover, the utility modeling function must also satisfy the completeness and transitivity properties. Due to the consideration of many variables in the behavioral modeling approach, the resulting mathematical formulation is not easily tractable. On the contrary, the axiomatic approach in rational choice theory results in a comparatively elegant mathematical framework. Therefore, rational choice theory is widely used to model the behavior of consumers and firms in an economy market [56]. Rational choice theory results in the consumer and firm theories to analyze economy markets. Consumer theory deals with the demand side and firm theory deals with the supply side of the market economy [53]. Cost and utility modelings are used in these theories to quantify and analyze the effects of decision-making by the two entities on the overall economy. In this context, cost and utility modeling has a crucial role in the microeconomics modeling and analysis.

1.4.2 Mathematical Modeling in Microeconomics

Mathematical methods provide sophisticated tools to model and analyze microeconomics models [57]. In particular, continuity, differential and convex theories are very handy in modeling, analyzing and verifying the cost and utility models in microeconomics [53]. Mathematical functions, such as polynomials, exponentials and logarithms, are widely used to model the cost and utility in smart grids.

Utility function is used to model the satisfaction behavior of a consumer in consuming goods or services. A mathematical function which is continuous, differentiable, non-decreasing and concave or strictly concave can be used to model the utility of a consumer or firm. The continuity and differentiability properties of the function ensure that the preferences of a consumer do not vanish abruptly over a given set of choices. Similarly, the non-decreasing property of the utility modeling function ensures that the choices and preferences are always intended to maximize the happiness or satisfaction of the individual. Moreover, the utility modeling function must also satisfy the completeness and transitivity properties. Completeness and transitivity mean that a consumer is always able to prefer and rank his choices for a given set of goods or services. Finally, the mathematical properties of concavity or strict concavity on the utility functions model another important notion in microeconomics which is known as the law of the diminishing marginal utility. This law states that every extra unit of the consumption leads to a reduced satisfaction or happiness of the consumer which in turn defines the upper bound on the utility.

The cost function is used to incorporate the total cost of producing a unit of goods or services. Cost of the goods or services is modeled using a continuous, differentiable, increasing and convex or strictly convex mathematical function. The product theory in microeconomics is used to model the production process of goods or services. In product theory, a product set, containing all possible feasible plans for production, is used to model the production process and it satisfies certain properties, such as non-emptiness, convexity and non-decreasing returns to the scale. The set relating the input and output of the feasible plans and this relationship is modeled using the cost function. The mathematical properties of monotonicity and convexity ensure that the production process follows the law of the marginal non-decreasing returns. The law describes a special relationship between the input and output of the production process, i.e., an addition of a production factor at the input, beyond the production capacity, results in smaller increase in the output. The condition of continuity on the modeling function ensures the existence of the solution for the corresponding cost minimization problem. Similarly, differentiability allows to define the notions of average and marginal costs for the production process.

Most of the mathematical properties of cost and utility functions can be modeled using the differential theory. In this regard, first-order and second-order derivatives are used to analyze the monotonicity and convexity or concavity of the given function, respectively [58]. The first-order derivative is also employed in the profit maximization problem to find the critical or stationary points of the function, i.e.,

$$\frac{\mathrm{d}f(x)}{\mathrm{d}x} = 0. \tag{1.3}$$

Equation (1.3) is also referred as the first-order condition [53]. The convexity or concavity property of a function is analyzed using the second-order derivative test. For a convex function, the second-order derivative condition is mathematically described as

$$0 \le \frac{\mathrm{d}^2 f(x)}{\mathrm{d}x^2}, \tag{1.4}$$

whereas for a concave function, second-order derivative is negative. For strictly convex and concave functions, the second-order derivatives are strictly positive and negative, respectively. However, converse of these conditions on the strict convex or concave functions are not valid. These conditions are mathematically expressed as strict inequalities, i.e.,

$$\text{strict-convexity} \Rightarrow 0 < \frac{\mathrm{d}^2 f(x)}{\mathrm{d}x^2}, \tag{1.5a}$$

$$\text{strict-convcavity} \Rightarrow \frac{\mathrm{d}^2 f(x)}{\mathrm{d}x^2} < 0. \tag{1.5b}$$

We use the mathematical framework of (1.3)–(1.5b), to develop a logical framework for the cost and utility models in Chap. 4 using the theorem-proving technique.

1.5 Algorithm Design

Many state-of-the-art operations of smart grids, such as scheduling algorithms for PEVs [27, 28, 36] and online pricing DR programs [27], involve decision-making algorithms using large data, such as history and population of consumers. Therefore, low computational complexity algorithms play a crucial role in the successful execution of such grid operations. In this regard, algorithm design theory provides mathematical basis for assessing the run-time and memory space bounds and requirements, respectively, for an algorithm. In this section, we introduce the computational complexity and asymptotic notations which are used in Chap. 5 to formally model computational complexity of smart grid algorithms for PEVs.

1.5.1 Computational Complexity

An algorithm is a procedure that gives the desired output to a given sequence of input [59]. Computers have bounded resources in terms of time and memory, and therefore it is mandatory for an algorithm to compute the given task in bounded time and memory space. Execution time and memory requirement for an algorithm are called the complexity of an algorithm. However, algorithms are just procedures to specify a logic to solve a given problem using computing capabilities of the machine. Practically, a given task may vary in its size, and therefore exact knowledge of the execution time and memory space cannot be obtained beforehand. Therefore, asymptotic notations [60] fit very naturally in the context of algorithm design to quantify the complexity of an algorithm. Asymptotic notations are mathematical tools to compare the growth rates of functions. Rates of growth allow us to distinguish or equate different functions on the basis of how fast a function increases with respect to the size of input. For running time analysis, one is usually interested in finding the number of computations and memory units, e.g., bytes, for computer memory space analysis. To analyze the asymptotic behavior of an algorithm, it is therefore an essential step to transform a given program into a mathematical equivalent function, as shown in Fig. 1.4. In this book, we focus on the computational complexity of algorithms and, therefore, we provide the detailed procedure to determine the computational complexity now.

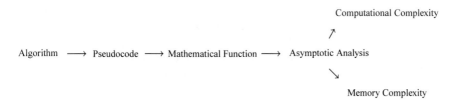

Computational Complexity

Algorithm ⟶ Pseudocode ⟶ Mathematical Function ⟶ Asymptotic Analysis

Memory Complexity

Fig. 1.4 Asymptotic analysis of an algorithm

Table 1.2 Asymptotic notations

Notation	Limit definition	Set definition
$f = O(g)$	$\lim\limits_{n\to\infty}\frac{f(n)}{g(n)} \neq \infty$	$\{f(n)|(\exists c\,n_o.\,(\forall n_o < n \wedge 0 < c \Rightarrow |f(n)| \leq cg(n)))\}$
$f = \Omega(g)$	$\lim\limits_{n\to\infty}\frac{f(n)}{g(n)} \neq 0$	$\{f(n)|(\exists c\,n_o.\,(\forall n_o < n \wedge 0 < c \Rightarrow cg(n) \leq |f(n)|))\}$
$f = \Theta(g)$	$\lim\limits_{n\to\infty}\frac{f(n)}{g(n)} \neq 0, \infty$	$\{f(n)|\exists c_1 c_2 n_o.\,(\forall n_o < n \wedge 0 < c_1 \wedge 0 < c_2 \Rightarrow$ $c_1 g(n) \leq |f(n)| \leq c_2 g(n)))\}$
$f = o(g)$	$\lim\limits_{n\to\infty}\frac{f(n)}{g(n)} = 0$	$\{f(n)|(\exists c\,n_o.\,(\forall n_o < n \wedge 0 < c \Rightarrow |f(n)| < cg(n)))\}$
$f = \omega(g)$	$\lim\limits_{n\to\infty}\frac{f(n)}{g(n)} = \infty$	$\{f(n)|(\exists c\,n_o.(\forall n_o < n \wedge 0 < c \Rightarrow cg(n) < |f(n)|))\}$

An algorithm uses primitive operations, such as addition and multiplication, to solve a problem [59]. The count of these basic operations is referred to as computational or run-time complexity of the algorithm. In order to count these steps, a mathematical function is obtained from the pseudocode of an algorithm. A pseudocode is a semi-formal description of an algorithm logic using a mix of natural and computer languages to express the logic of the algorithm. Natural language is used to briefly describe important information related to every step, and conventional programming language control constructs are used to describe mathematical or logical operations and recursions in a pseudocode. To obtain a mathematical function, the number of basic primitive operations are counted along with the cost of execution of each step. The asymptotic behavior [60] of these functions with respect to the size of input corresponds to the computational complexity of the given algorithm. These functions can be used to judge the best, worst and average case computational complexities of the algorithms. Table 1.2 lists some of the commonly used asymptotic notations which are used to characterize the computational efficiency of an algorithm and also provides metrics for comparative performance analysis among different algorithms [59].

In Table 1.2, Big-O is a set that contains all functions satisfying the condition, $f(n) \leq cg(n)$, whereas c is a positive constant and n is a natural number representing the size of the input, which should be greater than a specific threshold value, n_0. Similarly, Big-Ω represents a set that contains all functions satisfying the condition, $cg(n) \leq f(n)$. As the definitions suggest, the Big-O is an upper bound on the functions and hence captures the worst-case scenario of running time behavior of an algorithm, as shown in Fig. 1.5a. On the other hand, Big-Ω gives a lower bound on the set of given functions, as shown in Fig. 1.5b. In case of Big-Θ, the set members of the function have both upper and lower bounds, i.e., $c_1 g(n) \leq |f(n)| \leq c_2 g(n)$, where c_1 and c_2 are constant multipliers. Big-Θ notation is used to describe computational complexity of the algorithms that have the same best- or worst-case run-time. Lastly, little-o and little-ω sets have strict inequality conditions on the rate of the growths of the functions and hence are used to express weak lower and upper bounds on the computational complexity of the algorithms, respectively.

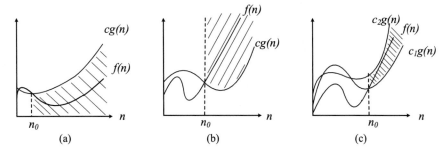

Fig. 1.5 Asymptotic bounds (from (**a**)–(**c**)). **a** Big-O notation; **b** Big-omega notation; **c** Big-theta notation

1.5.2 Online Scheduling Algorithms for Plug-in Electric Vehicles

Electrification of transportation technology is widely gaining popularity due to its low green-house emissions, energy security and noise mitigation features. Therefore, electric vehicles (EVs) are rapidly becoming a viable alternative to the conventional fossil-fuel-based transportation, e.g., more than 1 million electric vehicles were sold worldwide in the year 2017 alone [61] and the numbers are constantly increasing since then. One of the most important factors in the exponential growth of the electric vehicles is the concept of smart grids which provides an enabling environment for the large-scale integration of the electric vehicles [63, 64].

Plug-in electric vehicles are featured by an onboard battery storage, a charging plug and an internal combustion engine to charge the battery system and drive the vehicle [63]. In conjunction with the smart grid information and communication technologies, electric vehicles can act as an energy consumer and producer, also known as prosumer, i.e., from vehicle to the grid and from the grid to vehicle [64]. Smart grids leverage upon this characteristic of PEVs for load flattening, peak shaving and frequency fluctuation mitigation [66].

However, induction of a large number of PEV poses some serious challenges [65], e.g., network congestion, off-nominal frequency problems and three-phase voltage imbalance. In this regard, optimal PEV charging/discharging scheduling algorithms play a vital role in alleviating the problems arising from the large penetration of PEVs in smart grids.

Scheduling of PEVs is a decision-making problem, and therefore smart gird utilizes optimization techniques subject to the constraints imposed by the infrastructure and physical properties of the network and PEVs. Moreover, various approaches, such as linear programming [67], nonlinear programming [35], model predictive control [68] and queuing theory [69], are used to solve the optimization problem. The scheduling algorithms are categorized as online and offline scheduling algorithms depending on the nature of the information, i.e., causal or non-causal. An offline algorithm uses prior information about a PEV [70, 71], e.g., arrival time of

Fig. 1.6 Online PEV
scheduling algorithm

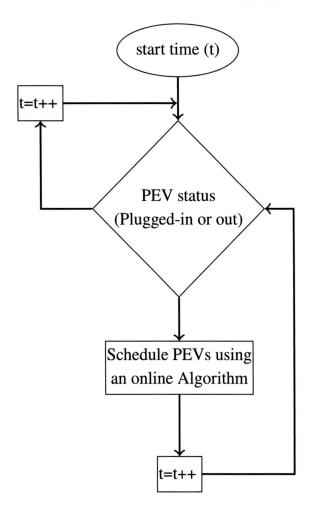

PEV, to schedule the charging or discharging of the PEV. However, this assumption fails under the real-world circumstances as a PEV can connect to the grid through a walled socket at any time. Moreover, there are many other factors that affect the optimal PEV scheduling, such as uncertainties arising from the renewable energy sources, intermittent power flow and varying energy demands and prices of electricity in the smart grid [66]. Therefore, online algorithms best suit to schedule PEV charging/discharging decisions as they only rely on the real-time information of network and PEV energy demands to make decisions, as shown in Fig. 1.6. This helps smart grids to ensure the efficient allocation of charging rates to grid connected PEVs and hence avoiding the problems arising from the penetration of large volumes of PEVs into the network, such as network congestions.

An online algorithm scheduler employs current information, called PEV profile, as well as the historical data and statistics to improve the efficiency of the algorithm

[36]. This approach is helpful in countering the uncertainty problems due to the integration of a large number of PEVs in the smart grids, however, low computational cost is direly needed to reap the benefits of online scheduling. Contrarily, computational complexity is a direct function of the population of PEVs. An online algorithm with high computational cost can jeopardize the smart grid objectives, such as reliability, efficiency and security. Therefore, recently many online low computational complexity scheduling algorithms have been proposed to ensure the optimal scheduling of PEVs [28, 68, 72].

The asymptotic notations, in Table 1.2, and pseudocode-based computational analysis of algorithms, described in this section, will be used to develop a logical framework in Chap. 5.

1.6 Traditional Analysis Techniques

Traditional analysis techniques, such as paper-and-pencil proof methods and simulations, are used to conduct the stability, cost and utility modeling and asymptotic analysis. Paper-and-pencil proof methods have been extensively used to conduct the analysis presented in Sects. 1.3, 1.4 and 1.5 based on foundational mathematical theories. Stability analysis [50] requires algebra, real and complex analysis to find roots of characteristic equations. Moreover, Routh-Hurwitz is a powerful paper-and-pencil method for the design of stable systems [50]. Similarly, the cost and utility modeling in smart grid applications heavily relies on the differential and convex theories. Finally, paper-and-pencil methods utilize the concepts of limits to analyze the asymptotic behavior of the functions. Computer-based simulations are also widely utilized for the analysis and verification of stability and cost and utility modeling in smart grids. Computer-based tools facilitate stability analysis using numerical methods [73]. General-purpose simulation tools, such as MATLAB [74], provide very useful packages to design and analyze the stability of control systems. Moreover, these tools also facilitate numerical methods-based differential analysis which is mainly employed in the cost and utility modeling [75]. Computer algebra systems [76] are another category of computer-based tools, which rely upon the symbolic manipulation, to analyze mathematical models, such as differential and root analysis for stability and microeconomics analysis.

Although the above-mentioned traditional analysis techniques are easy to use, these techniques do not provide an accurate and complete analysis of systems. Paper-and-pencil methods are prone to human error. Additionally, there is always a possibility that many assumptions and specifications are not documented due to human error or simply due to the intuitive nature of these assumptions and specifications. In case of large systems, it becomes almost impossible to keep track of all assumptions and specifications arising in many steps of the analysis. On the other hand, computer-based tools use numerical methods to execute given models which leads to unavoidable rounding, discretization and convergence errors [77], and therefore cannot model continuous behaviors of systems completely. Another problem with most of the computer-based tools is their sampling-based in-exhaustive analysis due to which several corner cases can be left undetected. Due to numerical methods and the sampling-based nature, computer-based tools cannot model continuous behaviors that are direly needed in the stability and microeconomics analysis. To alleviate the above-mentioned issues, computer algebra systems facilitate mathematical analysis based on the symbolic manipulation. However, these manipulations are based on large unverified computer programs which do not guarantee the coverage of all corner cases of the given problem [78]. Particularly, in case of asymptotic analysis, simulations and computer algebra systems offer very little help due to the involvement of abstract notions of limiting behavior of functions. Therefore, traditional techniques cannot ascertain an accurate analysis and verifications of smart grid problems involving concepts of stability, microeconomics and algorithm notations, whereas, in many safety- and mission-critical applications of smart grids, a minor bug can lead to irreparable loss, such as human life or financial loss. Therefore, in this chapter we propose formal methods to formally analyze and verify safety- and mission-critical aspects of smart grids.

1.7   Formal Methods

Formal methods [41] are computer-based mathematical techniques to model, specify and verify software and hardware systems. Formal methods are characterized as sound and complete analysis techniques. Soundness is assured by using well-defined syntax, semantics and proof theory of the underlying logic which ensures that only valid arguments are provable. On the other hand, the computer-based framework, along with the use of deductive logic, allows to exhaustively verify systems and hence it is ensured that there is no false positive, i.e., completeness property of the system.

Formal method techniques are an amalgam of logic, algebra, calculus and system modeling techniques within computers [83], as shown in Fig. 1.7. Hoar and Floyd were among the first to introduce logic within computers to reason about the correction of computer programs [85]. A constant pursuit for computer-based logical

Fig. 1.7 Formal methods

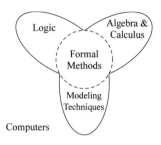

frameworks led to the higher-order logic, modal logic, temporal logic, etc., to express and reason about various abstractions of systems [84, 86]. Algebraic methods are used to embed structures in logics using operators and sets whereas, calculus is employed for symbolic manipulation of logical expressions [87, 88]. Lastly, system modeling techniques are used to conceptualize different kinds of systems encountered in nature, such as reactive systems. Finite state machines [89], Kripke structures [90] and binary decision trees [91] are well-known system modeling techniques, which are used in formal methods. A mix of these three major ingredients leads to the development of various formal methods for many engineering systems. There are two major formal method techniques, theorem proving and model checking [92]. These methods have been employed to formally verify and specify various software and hardware systems.

1.7.1 Model Checking

Model checking [92] is a process algebra-based formal methods technique. Process algebra is an approach, constituting specification languages and corresponding calculi, to formally verify and analyze concurrent systems. Temporal logic is one of the key features of model checking which enables the verification method to address time notion using logical operators, such as *eventually* and *until*. In model checking, a system is modeled as a finite state machine (FSM) and system's property is expressed using temporal logic. In this way, model checking exhaustively searches the state space of a given FSM for the property verification of the system. Model checking facilitates to formally verify safety and reachability properties of a systems. Therefore, these techniques are instrumental in formally verifying parallel and concurrent systems [93]. Computer programs which implement these formal methods are known as model checkers. A model checker returns true if property is valid and a counter example in case property is not valid with respect to the state space of a given FSM. This helps formal verification engineers to design and verify systems.

Due to the safety- and mission-critical nature of the smart grids, model checking has been employed to ensure safe and secure operations of these systems [94–97]. However, this technique may suffer from the state-explosion problem due to very large state space of the underlying system [98], such as continuous systems. Moreover, these techniques are not efficient at formally verifying functional models. Due to continuous nature of stability and microeconomics analysis and limiting behavior (a functional property) of asymptotic analysis, model checking cannot be used to accurately model and verify smart grids aspects involving these concepts.

1.7.2   Theorem Proving

Theorem proving is a formula-based formal methods approach [99]. The system that needs to be verified along with its requirement is expressed as a formula in logic. To formally verify that the system exhibits the given property, logical reasoning using proof theory of the corresponding logic is used. First-order, second-order and higher-order logic are used to formally describe systems and their properties. Computer programs which implement these formal method techniques are called theorem provers or proof assistants. Theorem proving is further classified as automatic and interactive. Automatic theorem provers [100] utilize decidable logics to formally express, reason and verify systems. But expressiveness is limited in decidable logics and therefore they cannot express and verify complex systems. On the other hand, interactive theorem [101] proving uses undecidable logics and therefore needs interaction with humans to reason and verify systems. However, these undecidable logics, such as higher-order logic, are highly expressive and therefore allow formally verifying continuous systems. Therefore, we utilize the interactive theorem-proving technique to formally verify safety- and mission-critical aspects of smart grids.

1.8   Methodology

Figure 1.8 describes the proposed theorem proving based methodology for the formal analysis and verification of key smart grid aspects which utilize stability, microeconomics and algorithm design analysis.

To conduct the formal stability analysis, we formally model stability criterion in higher-order logic as a set that contains all the roots lying in the left half s-plane and satisfying the characteristic polynomial represented by (1.2). To enable the formal verification using the formal stable criterion, we formally verify roots of polynomials of up to the fourth order within the sound core of a theorem prover. Finally, we conduct the formal stability analysis of power converter designs in smart grids using the proposed formalization.

To conduct formal analysis and verification of cost and utility models in microeconomics, we formally model cost and utility models using formal convex and differential theories. In order to formally reason about the properties of these models (described in Sect. 1.4.2), we formalize the derivative tests in higher-order logic. We, then, utilize the formalized derivative test and formal root analysis to accomplish the formal verification of cost and utility models. Based on the above formalization, we conduct the formal verification of fuel curve estimates of a thermal power plant.

Finally, to conduct the asymptotic analysis of algorithms, we, first, formally model asymptotic notations (given in Table 1.2), i.e., Big-O, Big-Ω, Big-Θ, Little-o and Little-ω, using the set theory. In order to conduct the formal analysis using these formal models, we also formally verify the properties of these asymptotic notations

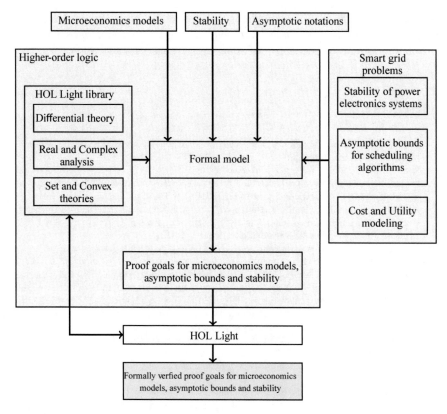

Fig. 1.8 Theorem proving based methodology for smart grids

within the sound core of a theorem prover. This formalization is then employed to formally verify the computational complexity bounds of online scheduling algorithms for plug-in electric vehicles.

We use the HOL Light [102] theorem prover, i.e., a software package, to develop the formalization due to its support for set, convex, differential, real and complex analysis theories, which are the foremost requirements for the proposed framework in this book.

1.9 Summary

This chapter provided a brief overview of smart grids and formal methods. Smart grid is a complex network of intelligent nodes with bidirectional flow of information and power which makes it challenging to verify these systems. Due to their safety- and mission-critical applications, we have proposed formal method techniques to formally verify smart grids. Formal methods are mathematical logic-based techniques with varying degrees of complexity. Theorem proving has been employed due to its capability to formally model and verify continuous behaviors of smart grids.

References

1. C. Hall, P. Tharakan, J. Hallock, C. Cleveland, M. Jefferson, Hydrocarbons and the evolution of human culture. Nature **426**(6964), 318–322 (2003)
2. P. Somavat, S. Jadhav, V. Namboodiri, Accounting for the energy consumption of personal computing including portable devices, in *Proceedings of the 1st International Conference on Energy-Efficient Computing and Networking* (2010), pp. 141–149
3. A. Shehabi, S. Smith, D. Sartor, R. Brown, M. Herrlin, J. Koomey, E. Masanet, N. Horner, I. Azevedo, W. Lintner, United states data center energy usage report (2016)
4. I.G.E. Outlook, Entering the decade of electric drive (2020)
5. J.A. Momoh, *Energy Processing and Smart Grid* (Wiley, New York, 2018)
6. S.S. Madani, E. Schaltz, K.S. Knudsen, An electrical equivalent circuit model of a lithium titanate oxide battery. Batteries **5**(1), 31 (2019). Multidisciplinary Digital Publishing Institute
7. S. Greene, I. Dobson, F.L. Alvarado, Sensitivity of the loading margin to voltage collapse with respect to arbitrary parameters. IEEE Trans. Power Syst. **12**(1), 262–272 (1997). IEEE
8. U.S Energy Information Administration.: Energy use in industry. https://www.eia.gov/energyexplained/use-of-energy/industry.php (2020). Accessed March 2021
9. L. Pérez-Lombard, J. Ortiz, C. Pout, A review on buildings energy consumption information. Energy Build. **40**(3), 394–398 (2008)
10. J.A. Momoh, *Smart Grid: Fundamentals of Design and Analysis*, vol. 63 (Wiley, New York, 2012)
11. V. Giordano, F. Gangale, G. Fulli, M.S. Jiménez, I. Onyeji, A. Colta, I. Papaioannou, A. Mengolini, C. Alecu, T. Ojala et al., Smart grid projects in Europe. JRC Ref. Rep. Sy. **8** (2011)
12. Q.C. Zhong, T. Hornik, *Control of Power Inverters in Renewable Energy and Smart Grid Integration*, vol. 97 (Wiley, New York, 2012)
13. Q. Wang, X. Liu, J. Du, F. Kong, Smart charging for electric vehicles: A survey from the algorithmic perspective. vol. 97. IEEE Commun. Surv. Tutorials **18**(2), 1500–1517 (2016)
14. D. Gan, D. Feng, J. Xie, *Electricity Markets and Power System Economics* (CRC Press, 2013)
15. P. Hasanpor Divshali, B.J. Choi, Electrical market management considering power system constraints in smart distribution grids. Energies **9**(6), 405 (2016)
16. N.S. Nafi, K. Ahmed, M.A. Gregory, M. Datta, A survey of smart grid architectures, applications, benefits and standardization. J. Netw. Comput. Appl. **76**, 23–36 (2016)
17. S. Srinivasan, Power relationships: Marginal cost pricing of electricity and social sustainability of renewable energy projects. Technol. Econ. Smart Grids Sustain. Energy **4**(1), 1–12 (2019)
18. S.S. Arefin, N. Das, Optimized hybrid wind-diesel energy system with feasibility analysis. Technol. Econ. Smart Grids Sustain. Energy **2**(1), 9 (2017)
19. A.N. Siddiqui, M.S. Thomas, Techno-economic evaluation of regulation service from sevs in smart mg system. Technol. Econ. Smart Grids Sustain. Energy **1**(1), 15 (2016)
20. W. Ketter, J. Collins, CA Block, Smart grid economics: Policy guidance through competitive simulation (2010)
21. M. Moretti, S.N. Djomo, H. Azadi, K. May, K. De Vos, S. Van Passel, N. Witters, A systematic review of environmental and economic impacts of smart grids. Renew. Sustain. Energy Rev. **68**, 888–898 (2017)
22. M. Greer, *Electricity Cost Modeling Calculations* (Academic, 2010)
23. R. Deng, Z. Yang, M.Y. Chow, J. Chen, A survey on demand response in smart grids: Mathematical models and approaches. IEEE Trans. Indus. Inf. **11**(3), 570–582 (2015)

24. S.M. Amin, B.F. Wollenberg, Toward a smart grid: power delivery for the 21st century. IEEE Power Energy Mag **3**(5), 34–41 (2005)
25. Y. He, B. Venkatesh, L. Guan, Optimal scheduling for charging and discharging of electric vehicles. IEEE Trans. Smart Grid **3**(3), 1095–1105 (2012)
26. F.Q. Igbinovia, J. Krupka, Computational complexity of algorithms for optimization of multi-hybrid renewable energy systems, in *2018 International Conference on Power System Technology (POWERCON)* (IEEE, 2012), pp. 4498–4505
27. P. Samadi, A.H. Mohsenian-Rad, R. Schober, V.W.S. Wong, J. Jatskevich, Optimal real-time pricing algorithm based on utility maximization for smart grid. IEEE Trans. Smart Grid 415–420 (2010). IEEE
28. W. Tang, Y.J.A. Zhang, A model predictive control approach for low-complexity electric vehicle charging scheduling: optimality and scalability. IEEE Trans. Power Syst. **32**(2), 1050–1063 (2016)
29. P. Samadi, A.H. Mohsenian-Rad, R. Schober, V.W.S. Wong, J. Jatskevich, Autonomous demand-side management based on game-theoretic energy consumption scheduling for the future smart grid. IEEE Trans. Smart Grid **1**(3), 320–331 (2010). IEEE
30. Z. Qiu, G. Deconinck, R. Belmans, A literature survey of optimal power flow problems in the electricity market context, in *2009 IEEE/PES Power Systems Conference and Exposition* (IEEE, 2009), pp. 1–6
31. Y. Sönmez, Estimation of fuel cost curve parameters for thermal power plants using the ABC algorithm. Turkish J. Electric. Eng. Comput. Sci. **21**(1), 1827–1841 (2013). IEEE
32. F.A. Wolak, Identification and estimation of cost functions using observed bid data: an application to electricity markets (2001)
33. M. Fahrioglu, F.L. Alvarado, Using utility information to calibrate customer demand management behavior models. IEEE Trans. Power Syst. **16**(2), 317–322 (2001)
34. H. Bessembinder, M. Lemmon, Equilibrium pricing and optimal hedging in electricity forward markets. J. Finance **57**(3), 1347–1382 (2002). Wiley Online Library
35. J. Soares, T. Sousa, H. Morais, Z. Vale, P. Faria, An optimal scheduling problem in distribution networks considering v2g, in *IEEE Symposium on Computational Intelligence Applications in Smart Grid (CIASG)* (IEEE, 2011), pp. 1–8
36. W. Tang, S. Bi, Y.J. Zhang, Online charging scheduling algorithms of electric vehicles in smart grid: An overview. IEEE Commun. Mag. **54**(12), 76–83 (2016)
37. U.S. Department of Energy (DOE), The Smart Grid: An Introduction. Available from internet (2008) https://www.energy.gov/sites/prod/files/oeprod/DocumentsandMedia/DOE_SG_Book_Single_Pages%281%29.pdf
38. Y. Yang, Y.E. Chen, Z. Liu, Energy constraints and China's economic development. J. Econ. Policy Reform **10**(4), 343–354 (2007). Taylor and Francis
39. A. Walker, E. Cox, J. Loughhead, J. Roberts, Counting the cost: the economic and social costs of electricity shortfalls in the UK-A report for the Council for Science and Technology. Royal Academy of Engineering (2014)
40. Y. Xu, C. Liu, K.P. Schneider, F.K. Tuffner, D.T. Ton, Microgrids for service restoration to critical load in a resilient distribution system. IEEE Trans. Smart Grid **9**(1), 426–437 (2016)
41. O. Hasan, S. Tahar, Formal verification methods, in *Encyclopedia of Information Science and Technology* ed. by M. Khosrow-Pour (IGI Global, 2015), pp. 7162–7170
42. F. Yao, A. Demers, S. Shenker, A scheduling model for reduced cpu energy, in *Proceedings 36th Annual Symposium on Foundations of Computer Science* (IEEE, 1995), pp. 374–382
43. Harrison, J.: Formal verification at intel. In: 18th Annual IEEE Symposium of Logic in Computer Science, 2003. Proceedings., pp. 45–54. IEEE (2003)
44. C. Newcombe, T. Rath, F. Zhang, B. Munteanu, M. Brooker, M. Deardeuff, Use of formal methods at amazon web services (2014). See http://research.microsoft.com/enus/um/people/lamport/tla/formal-methods-amazon.pdf
45. M. Bozzano, H. Bruintjes, A. Cimatti, J.P. Katoen, T. Noll, S. Tonetta, Formal methods for aerospace systems, in *Cyber-Physical System Design From an Architecture Analysis Viewpoint* (Springer, Berlin 2017), pp. 133–159

46. A. Fantechi, Twenty-five years of formal methods and railways: what next? in *International Conference on Software Engineering and Formal Methods* (Springer, Berlin, 2013), pp. 167–183

47. X. Fang, S. Misra, G. Xue, D. Yang, Smart grid-The new and improved power grid: A survey. IEEE Commun. Surv. Tutorials **14**(4), 944–980 (2011)

48. G. Rietveld, J.P. Braun, R. Martin, P. Wright, W. Heins, N. Ell, P. Clarkson, N. Zisky, Measurement infrastructure to support the reliable operation of smart electrical grids. IEEE Trans. Instrum. Meas. **64**(6), 1355–1363 (2015)

49. E. Kabalci, Y. Kabalci, *Smart Grids and Their Communication Systems* (Springer, Berlin, 2019)

50. N.S. Nise, *Control Systems Engineering* (Wiley, New York, 2007)

51. P.P. Dyke, *An Introduction to Laplace Transforms and Fourier Series* (Springer, Berlin, 2014)

52. S. Skogestad, I. Postlethwaite, *Multivariable Feedback Control: Analysis and Design*, vol. 2 (Wiley, New York, 2007)

53. A. Mas-Colell, M.D. Whinston, J.R. Green et al., *Microeconomic Theory*, vol. 1 (Oxford University Press, New York, 1995)

54. N. Wilkinson, M. Klaes, An Introduction to Behavioral Economics (Macmillan International Higher Education, 2017)

55. J.S. Coleman, T.J. Fararo, *Rational Choice Theory* (Sage, Nueva York, 1992)

56. J. Levin, P. Milgrom, *Introduction to Choice Theory*. Available from internet. (2004) https://web.stanford.edu/~jdlevin/Econ%20202/Choice%20Theory.pdf

57. J.M. Henderson, R.E. Quandt, et al., *Microeconomic Theory: A Mathematical Approach* (1971)

58. D. Zill, W.S. Wright, M.R. Cullen, *Advanced Engineering Mathematics* (Jones AND Bartlett Learning, 2011)

59. T.H. Cormen, C.E. Leiserson, R.L. Rivest, C. Stein, *Introduction to Algorithms*, 2nd edn. (The MIT Press, 2001). http://web.ist.utl.pt/fabio.ferreira/material/asa/clrs.pdf

60. R.L. Graham, D.E. Knuth, O. Patashnik, *Concrete Mathematics: A Foundation for Computer Science*, 2nd edn. (Addison-Wesley Longman Publishing Co., Inc., 1994)

61. Cazzola, P., et al, M.G.: Global EV Outlook 2018 Towards cross-modal electrification (2018)

62. M. Pasquier, J.M. Mintz et al., IEA-HEV-TCP Task 24: Economic Impact Assessment of E-mobility (2016)

63. J.A.P. Lopes, F.J. Soares, P.M.R. Almeida, Integration of electric vehicles in the electric power system. Proc. IEEE **99**(1), 168–183 (2011)

64. C. Guille, G. Gross, A conceptual framework for the vehicle-to-grid (v2g) implementation. Energy Policy **37**(11), 4379–4390 (2009)

65. K. Clement-Nyns, E. Haesen, J. Driesen, The impact of vehicle-to-grid on the distribution grid. Electric Power Syst. Res. **81**(1), 185–192 (2011)

66. W. Tang, Y. Jun, *Optimal Charging Control of Electric Vehicles in Smart Grids* (Springer, Berlin, 2017)

67. C. Kurucz, D. Brandt, S. Sim, A linear programming model for reducing system peak through customer load control programs. IEEE Trans. Power Syst. **11**(4), 1817–1824 (1996)

68. W. Tang, S. Bi, Y.J.A. Zhang, Online coordinated charging decision algorithm for electric vehicles without future information. IEEE Trans. Smart Grid **5**(6), 2810–2824 (2014)

69. T. Zhang, W. Chen, Z. Han, Z. Cao, Charging scheduling of electric vehicles with local renewable energy under uncertain electric vehicle arrival and grid power price. IEEE Trans. Vehicular Technol. **63**(6), 2600–2612 (2014)

70. Z. Ma, D.S. Callaway, I.A. Hiskens, Decentralized charging control of large populations of plug-in electric vehicles. IEEE Trans. Control Syst. Technol. **21**(1), 67–78 (2013)

71. J. Aghaei, M.I. Alizadeh, Demand response in smart electricity grids equipped with renewable energy sources: A review. Renew. Sustain. Energy Rev. **18**, 64–72 (2013)

72. E.H. Gerding, V. Robu, S. Stein, D.C. Parkes, A. Rogers, N.R. Jennings, Online mechanism design for electric vehicle charging, in *The 10th International Conference on Autonomous Agents and Multiagent Systems*, vol. 2 (International Foundation for Autonomous Agents and Multiagent Systems, 2011), pp. 811–818

73. J.F. Epperson, *An Introduction to Numerical Methods and Analysis* (Wiley, New York, 2013)
74. MathWorks: Control System Toolbox. https://ch.mathworks.com/products/control.html (2006). Accessed 27 Feb 2021
75. K.L. Judd, K.L. Judd, Numerical Methods in Economics. MIT press (1998)
76. J. Von Zur Gathen, J. Gerhard, *Modern Computer Algebra* (Cambridge University Press, Cambridge, 2013)
77. J.M. Ortega, *Numerical Analysis: A Second Course* (SIAM, 1990)
78. D.R. Stoutemyer, Crimes and misdemeanors in the computer algebra trade. Notices Am. Math. Soc. **38**(7), 778–785 (1991)
79. T. Wei, S. Yang, J.C. Moore, P. Shi, X. Cui, Q. Duan, B. Xu, Y. Dai, W. Yuan, X. Wei et al., Developed and developing world responsibilities for historical climate change and co2 mitigation. Proc. Natl. Acad. Sci. **109**(32), 12911–12915 (2012)
80. R.G. Pratt, P.J. Balducci, C. Gerkensmeyer, S. Katipamula, M.C. Kintner-Meyer, T.F. Sanquist, K.P. Schneider, T.J. Secrest, The smart grid: an estimation of the energy and co2 benefits (2010)
81. W. Su, H. Eichi, W. Zeng, M.Y. Chow, A survey on the electrification of transportation in a smart grid environment. IEEE Trans. Indus. Inf. **8**(1), 1–10 (2011)
82. W. Strielkowski, Social and economic implications for the smart grids of the future. Econ. Soc. **10**(1), 310 (2017)
83. J. Qadir, O. Hasan, Applying formal methods to networking: theory, techniques, and applications. IEEE Commun. Surv. Tutorials **17**(1), 256–291 (2015)
84. M. Huth, M. Ryan, *Logic in Computer Science: Modelling and Reasoning About Systems* (Cambridge University Press, Cambridge, 2004)
85. C.A.R. Hoare, An axiomatic basis for computer programming. Commun. ACM **12**(10), 576–580 (1969)
86. A. Galton, Logic as a formal method. Comput. J. **35**(5), 431–440 (1992)
87. D. Kozen, Kleene algebra with tests. ACM Trans. Program. Lang. Syst. (TOPLAS) **19**(3), 427–443 (1997)
88. J.C. Baeten, T. Basten, T. Basten, M. Reniers, *Process Algebra: Equational Theories of Communicating Processes*, vol. 50 (Cambridge University Press, Cambridge, 2010)
89. P. Salem, Practical programming, validation and verification with finite-state machines: a library and its industrial application, in *Proceedings of the 38th International Conference on Software Engineering Companion* (2016), pp. 51–60
90. S. Chaki, E.M. Clarke, J. Ouaknine, N. Sharygina, N. Sinha, State/event-based software model checking, in *International Conference on Integrated Formal Methods* (Springer, Berlin, 2004), pp. 128–147
91. C. Appold, Improving bdd based symbolic model checking with isomorphism exploiting transition relations (2011). arXiv:1106.1229
92. E.M. Clarke, O. Grumberg, D.E. Long, Model checking and abstraction. ACM Transactions on Programming Languages and Systems (TOPLAS) **16**(5), 1512–1542 (1994)
93. D. Bošnački, A. Wijs, *Model Checking: Recent Improvements and Applications* (Springer, Berlin, 2018)
94. A. Mahmood, O. Hasan, H.R. Gillani, Y. Saleem, S.R. Hasan, Formal reliability analysis of protective systems in smart grids, in *2016 IEEE Region 10 Symposium (TENSYMP)* (IEEE, 2016), pp. 198–202
95. A. David, D. Du, K.G. Larsen, M. Mikučionis, A. Skou, An evaluation framework for energy aware buildings using statistical model checking. Sci. China Inf. Sci. **55**(12), 157–168 (2012). Springer
96. S. Patil, G. Zhabelova, V. Vyatkin, B. McMillin, Towards formal verification of smart grid distributed intelligence: Freedm case, in *IECON 2015-41st Annual Conference of the IEEE Industrial Electronics Society November 2015* (IEEE, 2015), pp. 003974–003979
97. Y. Kwon, E. Kim, S. Jeong, A.H. Lee, Quantitative model checking for a smart grid pricing, in *International Conference on Quantitative Evaluation of Systems 2017* (Springer, Cham, 2015), pp. 55–71

98. E.M. Clarke, W. Klieber, M. Nováček, P. Zuliani, Model checking and the state explosion problem, in *LASER Summer School on Software Engineering* (Springer, Berlin, 2011), pp. 1–30
99. J. Harrison, J. Urban, F. Wiedijk, History of interactive theorem proving. Comput. Logic **9**, 135–214 (2014)
100. J. Harrison, A short survey of automated reasoning, in *International Conference on Algebraic Biology* (Springer, Berlin, 2007), pp. 334—49
101. F. Maric, A survey of interactive theorem proving. Zbornik radova **18**(26), 173–223 (2015)
102. J. Harrison, HOL-Light Theorem Prover. https://www.cl.cam.ac.uk/~jrh13/hol-light/. Accessed 1 March 2021

Chapter 2
Interactive Theorem Proving

Traditional analysis techniques cannot ascertain an accurate analysis for the safety- and mission-critical applications of smart grids due to their inability to exhaustively specify and verify the systems. To overcome these issues, we propose the theorem proving based methodology for an accurate analysis of smart grid problems which can be modeled using stability, microeconomics and algorithm design theories. We choose the HOL Light theorem prover to develop the proposed formalizations due to the availability of foundational theories, such as set, differential, real and complex, in the HOL Light library. In this chapter, we present a brief overview of the HOL Light theorem prover and commonly used formalizations from the HOL Light library to facilitate the understanding of the proposed formalizations which are presented, in the rest of the book.

2.1 Introduction

Theorem proving is categorized as automated theorem proving (ATP) and interactive theorem proving (ITP) [1]. The former is best suited for the systems which can be expressed using first-order logic, however, many mathematical concepts, such as continuity and differentiability, cannot be modeled using first-order logic. Therefore, ITP is adopted, which utilizes second- or higher-order logic to express more abstract mathematical concepts, along with the well-defined syntax, semantics and proof theory, to formally reason and verify mathematical models. Currently, many theorem provers are available, such as HOL Light [2], Isabelle [3], Lean [4] and PVS [5]. However, we opted for the HOL Light theorem prover mainly due to its large library of formal proofs from set, convex, differential, real and convex theories.

© The Author(s), under exclusive license to Springer Nature Switzerland AG 2022   23
A. Ahmed et al., *Formal Analysis of Future Energy Systems Using Interactive Theorem Proving*, SpringerBriefs in Applied Sciences and Technology,
https://doi.org/10.1007/978-3-030-78409-6_2

The rest of the chapter is organized as follows: Sect. 2.2 provides a brief overview of HOL Light and Sect. 2.3 provides higher-order logic modeling and mechanized proofs, from set and differential theory, to facilitate the formalizations presented in Chaps. 3–5. Finally, Sect. 2.4 concludes the chapter.

2.2   HOL Light Theorem Prover

HOL Light [2] is an interactive theorem prover, which is a descendant of (Logic Computable Functions) LCF style theorem provers [6]. HOL Light is developed by John Harrinson. In HOL Light, higher-order logic is implemented using Objective Camel (OCaml), which is a functional programming language. HOL Light is a strongly typed system, which contributes to the soundness of embedded logic. Basic building blocks in HOL Light are three objects: `hol_type`, `term` and `thm` [7]. `hol_type` represents types of the HOL Light objects, i.e., `term` and `thm`. `term` has a well-defined type which is used to express logical or mathematical assertions. New data types, `thm`, are obtained by applying higher-order-logic deduction rules on terms. HOL Light is featured for its minimalist development approach: relying on just 10 inference rules, with only 1500 lines of code in its core [8]. All new theorems (mathematical or logical formulas) have to be verified based on these 10 inference rules to preserve soundness. Typed λ-calculus [9] is used for the symbolic reductions. To ease the programming within the system, the system allows deriving logical inference rules and tactics without expanding the power of the system. Tactics are programs in OCaml, which allow conducting forward and backward proof styles and simplification rules (arithmetic and set theoretic), such as `MESON_TAC` and `ARITH_TAC` [10].

HOL Light is endowed with a large library of formalized mathematical theories, such as multivariate real and complex theories, which can be readily utilized for the formal analysis and verification of the safety- or mission-critical aspects of systems. HOL Light has been utilized, at intel, for the formal verification of the floating numbers [11]. HOL Light theorem prover has, also, been employed in the formal verification of Kepler's conjunction [12]. One of the criteria for the efficiency and performance of any theorem prover is the number of formally verified theorems from the list of "top 100" theorems [13]. HOL Light is on the top by formally verifying 86 theorems from the list [13]. Recently, due to higher-order logic and the large library of formal proofs, HOL Light has been recently used for the realization of general-purpose auto-formalization of proofs using state-of-the-art machine learning algorithms [14–16].

Table 2.1 presents a list of commonly used HOL Light functions and symbols to facilitate the understanding of the formalizations presented in this book.

Table 2.1 Higher-order-logic (HOL) symbols and functions

HOL Symbol	Standard Symbol	Meaning
$\wedge$	*and*	Logical *and*
$\vee$	*or*	Logical *or*
$\neg$	*not*	Logical *negation*
$\Longrightarrow$	$\rightarrow$	Logical *conditional*
%	$c\,\vec{a}$	Scalar multiplication of vectors
λx.t	$\lambda x.t$	Function that maps x to $t(x)$
num	$\{0, 1, 2, \ldots\}$	Positive integers data type
real	All real numbers	Real data type
complex	All complex numbers	Complex data type
SUC n	$n + 1$	Successor of a *num*
rpow x y	x^y	Power with real exponent
max (f(x), g(x))	$max(g(x), f(x))$	Maximum among $f(x)$ and $g(x)$

2.3 Formalized Mathematical Theories in HOL Light

We present basic definitions and properties of set, differential and convex theories of HOL Light that will be helpful in understanding the formalizations, presented in the next chapters of this book.

2.3.1 Set Theory

The set theory has a central role in the conception and development of the HOL Light theorem prover [17]. We present a few set operations and theorems that are used in our formal verification of smart girds.

A set membership ($\in$) operation in HOL Light is defined as

Definition 2.1 $\vdash \forall$ P x. x IN P = P x

Where P: $A \rightarrow$ *bool* is a predicate modeling a set.

An empty set is a modeled as follows:

Definition 2.2 $\vdash$ {} = (λx. F)

There are a number of theorems, which are used to eliminate set abstraction. We present the basic one which has been used in our formalizations,

Definition 2.3 $\vdash \forall$p. GSPEC p = p

Finally, a non-empty set-related theorem is as follows:

Theorem 2.1 $\vdash \forall x. \sim (x \text{ IN } \{\})$

The above theorem ensures that the empty set has no member.

In this book, the HOL Light library for sets has been extensively utilized to formally model stability and asymptotic notations in Chaps. 3 and 5.

2.3.2 Multivariate Theory

Multivariate theory of HOL Light provides reasoning support for topology, analysis and geometry in Euclidean space. Using these fundamentals, real, complex, integral and differential theories are formalized that have been used in the formal analysis and verification of several safety- and mission-critical aspects of smart grids. HOL Light utilized R^N data type to formally model N-dimensional Euclidean space [18]. An N-dimensional vector is then defined using the lambda function (which is a mapping function),

Definition 2.4 $\vdash \forall l. \text{ vector } l = (\text{lambda } i. \text{ EL } (i - 1) \ l)$

In the above definition, `vector` is a higher-order logic function, which accepts a list, l, and maps the elements to a R^N space. This ingenious approach allows to not only formally reason in finite N-dimensional space but also supports formal reasoning in subspaces, such as *real* and *complex*, by instantiating $N = 1$ and 2, respectively.

There are HOL Light functions which allow to map *real*, *complex* and *vector* spaces. A term of type *complex* in HOL Light is defined using Definition 2.4 as:

Definition 2.5 $\vdash \forall x \ y. \text{ complex}(x, y) = \text{vector } [x; y]$

A term of type real can be mapped as complex number using Cx,

Definition 2.6 $\vdash \forall a. \text{ Cx } a = \text{complex}(a, \&0)$

A term of type real[1] can be mapped to a real type using drop,

Definition 2.7 $\vdash \forall a. \text{ Cx } a = \text{complex}(a, \&0)$

Definition 2.8 $\vdash \forall x. \text{ drop } x = x\$1$

where $ represents dimension index. Whereas, the `lift` function in HOL Light is used to map a term of type real to real[1],

Definition 2.9 $\vdash \forall x. \text{ lift } x = (\text{lambda } i. \ x)$

The traditional Frechet derivative definition in HOL Light is expressed as:

Definition 2.10 $\vdash \forall f \ f' \ \text{net}.$
```
(f has_derivative f') net =
A1:linear f' ∧
((λy. inv (norm (y - netlimit net)))) %
(f y - (f (netlimit net) + f ' (y - netlimit net))) →
vec 0
```

In the above definition, `has_derivative` is a higher-order logic function, which accepts function $f: R^M \to R^N$, derivative of function $f_\prime : R^M \to R^N$ and evaluation point or interval `net : A`. The net has an arbitrary type, A, specified by the user. In Assumption `A1`, `linear` is another higher-order-logic predicate that ensures that the derivative function is a linear transformation. The conclusion of the above definition is a predicate ensuring that the limit of the derivative approaches zero.

Next, we present a few theorems from the Multivariate real analysis theory, which are utilized in the cost and utility modeling of smart grids. A real convex function is defined in higher-order logic as:

Definition 2.11 $\vdash \forall$ s f. f `real_convex_on` s = ($\forall$ x y u v. x $\in$ s $\land$ y $\in$ s $\land$ &0 $\leq$ u $\land$ &0 $\leq$ v $\land$ u + v = &1 $\Rightarrow$ f (u * x + v * y) $\leq$ u * f x + v * f y)

In Definition 2.11, `real_convex_on` is a higher-order logic function that accepts a real-valued function, $f: (\mathbb{R} \to \mathbb{R})$, defined over an interval s : real $\to$ bool, whereas s is a set theoretic definition of the real interval that represents all subintervals possible for a given real interval. The function is convex if for any two points x and y, within the real interval s, the value of the function f (u * x + u * y) lies below the line segment defined by the function values at f (x) and f (y). The variables u and v, of real data type, together provide all possible lines for the graph of the function between f (x) and f (y).

Based on the definition of `real_convex_on`, the following properties of convex functions have been formally verified in HOL Light:

Theorem 2.2 $\vdash \forall$ s c. (λx. c) `real_convex_on` s

Theorem 2.3 $\vdash \forall$ s f g.
A1 : f `real_convex_on` s $\land$ A2 : g `real_convex_on` s $\Longrightarrow$
(λx. f x + g x) `real_convex_on` s

Theorem 2.4 $\vdash \forall$ s c f.
A1 : 0 <= c A2 : $\land$ f `real_convex_on` s $\Longrightarrow$ (λx. c * f x) `real_convex_on` s

The second-order derivative test is widely used to analytically verify the convexity or concavity of a given function. The second-order test for the convexity of a function has been verified in the multivariate real analysis theory of HOL Light as

Theorem 2.5 $\vdash \forall$ f f$^\prime$ f$^{\prime\prime}$ s .
A1 : ¬($\exists$a. s = { a }) $\land$
A2 : ($\forall$x. x $\in$ s $\Rightarrow$ (f `has_vector_derivative` f$^\prime$ x) (atreal x within x))
A3 : ($\forall$x. x $\in$ s $\Rightarrow$ (f$^\prime$ `has_vector_derivative` f$^{\prime\prime}$ x) (atreal x within x))
$\Rightarrow$ f `real_convex_on` s = ($\forall$x. x $\in$ s $\Rightarrow$ &0 $\leq$ f$^{\prime\prime}$)

In the above theorem, f, f$^\prime$ and f$^{\prime\prime}$ represent a real-valued function, and its first-order and second-order derivatives, respectively, defined over an arbitrary interval, s. The first Assumption `A1` excludes all real subintervals, of a given interval, with a single element to satisfy the differentiability of the given function subject to the open intervals. Assumptions `A1`–`A2` ensure that the first-order and second-order derivatives of the given function exist within the given interval s, whereas `has_vector_derivative` is the definition of the real derivative in HOL Light in the relational form. Finally, the theorem concludes the equivalence of the convexity of a given function and its positive second-order derivative, i.e., $0 \leq f^{\prime\prime}$.

We present one more fundamental theorem of differential theory, i.e., mean value theorem, from the HOL Light multivariate library. Mechanized proof of mean value theorem, for real-valued functions, is available in HOL Light as

Theorem 2.6 $\vdash \forall$ f f$'$ a b.
A1 : drop a < drop b $\wedge$
A2 : (($\forall$x. x IN [a, b]
$\implies$ (f has_derivative f$'$ x) (at x wihtin interval [a, b])) $\implies$
($\exists$x. x IN interval (a, b) $\wedge$ f b $-$ f a = f$'$ x (a, b))

The mean value theorem has been employed to formally verify the strict real convex property of a given function in Chap. 4. Moreover, the proposed formalizations for stability, microeconomics models and asymptotic notations are mainly built in a complex domain with the help of the multivariate theory in HOL Light.

2.4 Summary

This chapter provided a brief overview of the HOL Light theorem prover and its libraries: set and differential theories. Formal definitions and theorems of sets, derivatives and real convex are presented that are mainly used in the formal modeling and analysis of smart grids in the next three chapters.

References

1. J. Harrison, J. Urban, F. Wiedijk, History of interactive theorem proving. Comput. Logic **9**, 135–214 (2014)
2. J. Harrison, HOL-Light Theorem Prover. https://www.cl.cam.ac.uk/~jrh13/hol-light/. Accessed 1 Mar 2021
3. Isabelle. https://isabelle.in.tum.de/. Accessed 1 Mar 2021
4. LEAN. https://leanprover.github.io/. Accessed 1 Mar 2021
5. PVS. https://pvs.csl.sri.com/. Accessed 1 Mar 2021
6. M. Gordon, From LCF to HOL: a short history, in *Proof, Language, and Interaction* (2000), pp. 169–186
7. J. Harrison, HOL Light tutorial. http://www.cl.cam.ac.uk/~jrh13/hol-light/tutorial.pdf. Accessed 1 Mar 2021
8. S. McLaughlin, C. Barrett, Y. Ge, Cooperating theorem provers: A case study combining HOL-Light and CVC Lite. Electron. Notes Theor. Comput. Sci. **144**(2), 43–51 (2006)
9. L.C. Paulson, Formalising mathematics in simple type theory, in *Reflections on the Foundations of Mathematics* (Springer, Berlin, 2019), pp. 437–453
10. J. Harrison, HOL Light Reference Manual. https://www.cl.cam.ac.uk/~jrh13/hol-light/reference.html. Accessed 1 Mar 2021
11. J. Harrison, Floating point verification in HOL Light: the exponential function, in *International Conference on Algebraic Methodology and Software Technology* (Springer, Berlin, 1997), pp. 246–260
12. T. Hales Flyspeck. https://code.google.com/archive/p/flyspeck/. Accessed 1 Mar 2021
13. J. Harrison Formalizing 100 theorems. http://www.cs.ru.nl/~freek/100/. Accessed 1 Mar 2021
14. A. Paliwal, S.M. Loos, M.N. Rabe, K. Bansal, C. Szegedy, Graph representations for higher-order logic and theorem proving, in *AAAI* (2020), pp. 2967–2974
15. K. Bansal, S. Loos, M. Rabe, C. Szegedy, S. Wilcox, Holist: An environment for machine learning of higher order logic theorem proving, in *International Conference on Machine Learning* (2019), pp. 454–463
16. C. Kaliszyk, J. Urban, Hol (y) hammer: Online atp service for hol light. Math. Comput. Sci. **9**(1), 5–22 (2015)

17. T.C. Hales, Developments in formal proofs (2014). arXiv:1408.6474
18. J. Harrison, The HOL light theory of euclidean space. J. Autom. Reason. **50**(2), 173–190 (2013)

Chapter 3
Formalization of Stability Theory

In smart grids, power electronics circuits are used to process power from distributed energy resources (DERs). Stability analysis is mandatory for the design of power electronics circuits. Stability analysis foundations, such as root analysis, are also used in load variance analysis and cost and utility function modeling in smart grids. Traditionally, paper-and-pencil proof methods and simulations are used to conduct stability analysis. But traditional techniques cannot guarantee accurate analysis of the systems due to their inherent limitations. However, an accurate and exhaustive stability analysis is direly needed in the safety- and mission-critical operations of smart grids, such as integration of electric vehicles and renewable energy sources. Therefore, in this chapter, we present a logical framework for the formal stability analysis using formal methods technique, i.e., theorem proving. The proposed formalization is, then, employed to formally verify the design of power electronics circuits for wind turbines.

3.1 Introduction

In this chapter, we present a formalization for the stability analysis of linear time-invariant control systems, represented by characteristic equations of order upto four, with minimal dependence on conventional analysis techniques. We consider complex polynomials with real coefficients, for the purpose of formal analysis in higher-order logic, which allow us to express the cubic and quartic complex polynomials in terms of the quadratic polynomials. However, this choice does not limit the scope of the applicability of our formalization, as these coefficients are usually real numbers as they represent the different parameters of the system, e.g., resistance in electrical and electronics systems. The formally verified roots, which are poles of the system, are then formally analyzed to check for the stability condition, i.e., if they lie in the

A. Ahmed et al., *Formal Analysis of Future Energy Systems Using Interactive Theorem Proving*, SpringerBriefs in Applied Sciences and Technology, https://doi.org/10.1007/978-3-030-78409-6_3

left-half of the complex-plane, in the sound core of the higher-order-logic theorem prover HOL Light [1]. The main motivation of this choice is the extensive reasoning support available in HOL Light about multivariate complex, real and transcendental theories, which are required for the formalization of stability analysis of control systems.

The rest of the chapter is organized as follows: In Sect. 3.2, we present a review of the related work. This is followed by the description of the proposed methodology about stability analysis in Sect. 3.3. The formalization of the quadratic, cubic and quartic characteristic polynomials is described in Sect. 3.4. We utilize this formalization to formally verify voltage and current controllers designed for the power converters for reliable and efficient smart grid operation in Sect. 3.5. Finally, Sect. 3.6 concludes the chapter.

3.2 Related Work

Formal methods have been employed to formally analyze and verify control systems. Laplace transform [2] is formalized to formally verify the time-domain representation of systems to frequency-domain representation. Important features of this formalization include formal verification of transform properties of existence, linearity and frequency shifting. This existing formalization can be used in conjunction with the proposed formal stability analysis framework to formally analyze the systems in the frequency domain. The work, presented in [3, 4], allows to formally verify the Block diagram technique which is used to obtain transfer functions from the interconnected subsystems of a control system, in open loop or close loop configurations. On the other hand, the formalization proposed in this chapter deals with the stability analysis which is based on the transfer function of the system. Stability analysis plays a very important role in the analysis of safety and mission-critical systems. A partial treatment of stability can be found in some formalizations proposed for safety or mission-critical applications. A formal stability analysis of optical waveguides [5] is conducted in terms of the parameters of waveguide system, such as orientation of a ray in a wave guide using HOL Light theorem prover. A platoon vehicle controller is designed and implemented in higher-order logic to perform runtime monitoring for a platoon of vehicles. A comprehensive formalization has been proposed to cover the formal analysis of control systems including frequency transformation, block diagram analysis and controller designs for the systems [6]. However, the aforementioned formalizations address stability of specific systems and are not based on a generic definition of stability. Finally, the formal analysis of cyber-physical system uses formal factorization of quadratic polynomial in the HOL4 theorem prover [8]. However, the factorization theorem for the quadratic polynomial in this work is handled using the real number theory of HOL4 library. Therefore, it is not possible to use it for formal stability analysis in the complex domain.

The formal stability analysis, proposed in this chapter, is unique, compared to the related work, due to the incorporation of stability definition in complex domain and formally verifying the conditions of stability for characteristic polynomials upto the fourth order. Moreover, the formal verification of factorization of the polynomials ensures that every root is considered for stability. This results in an exhaustive list of assumptions on the roots of the polynomials which are highly desirable to formally verify the stability of safety or mission-critical systems.

3.3 Proposed Methodology

We propose a theorem, proving based methodology, shown in Fig. 3.1, to formally model and verify stability of systems using the corresponding characteristic equations of systems

- As a first step, we formally model the stability condition for a root/eigen value of the characteristic polynomial in higher-order logic which enables us to formally specify and reason about the stability of a system. We formally define a set, using set theory of HOL Light, with members satisfying the stability conditions, as described in Sect. 1.3.

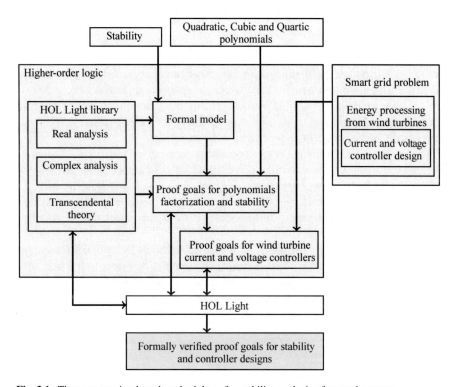

Fig. 3.1 Theorem proving based methodology for stability analysis of control systems

- In the next step, we formally verify the factorization of polynomials upto the fourth order to facilitate formal reasoning for the stability conditions on the roots of the given polynomial.
- In the third step, we use the above formalizations to mechanically verify all the conditions on the roots of the polynomials upto the fourth order using the sound core of HOL Light theorem.
- Finally, we utilize the proposed formalization to formally verify the current and voltage controllers for power converters of wind turbines in a smart grid.

We use the HOL light theories of real, complexes and transcendentals to formally reason about the stability conditions of the roots of the polynomials.

3.4 Stability Formalization

In this section, we present the higher-order logic theorem proving based formalization for stability theory. We formally model the stability condition and formally verify the factorization theorems for characteristic Eq. (1.2), upto the fourth degree to formally reason about their stability conditions in the sound core of HOL Light theorem prover.

The stability of a root is defined, as a higher-order logic function, using (1.2) as

Definition 3.1 $\vdash \forall$ f.
stable f = ~({ x | f x = Cx (0) $\wedge$ Re (x) < 0 }= EMPTY)

Definition 3.1 models the stability condition using set theory of HOL Light library. stable is a higher-order-logic function which accepts a complex function, $f: R^2 \rightarrow R^2$, i.e., characteristic polynomial, and uses Re and Cx functions to extract real and imaginary parts of the root, x, of the characteristic equation, respectively. Whereas, EMPTY is the higher-order logic function for the empty set.

The predicate stable imposes two conditions on the given polynomial function. The first condition, i.e., f x = Cx (0), requires that x is the root of the given polynomial to qualify the stable set membership, and the second condition, i.e., Re (x) < 0, ensures that the root satisfies the stability condition. Thus, the predicate ensures that for a stable system the roots of the characteristic equation satisfies the stability condition.

However, to formally reason about the stability conditions, Definition 3.1, requires the factorization of the given characteristic polynomial. Therefore, in the next sections, we formally verify the factor decomposition of polynomials upto the fourth order.

3.4.1 Quadratic Polynomial

We formally verify the factorization of the quadratic polynomial in HOL Light, as

Theorem 3.1 $\vdash \forall$ a b c x.
A1 : a $\neq$ 0
$\Rightarrow$ Cx a * x pow 2 + Cx b * x + Cx c = Cx 0
$$x = \frac{-\text{ Cx b} + \sqrt{\text{Cx b pow 2} - \text{Cx 4} * \text{Cx a} * \text{Cx c}}}{\text{Cx 2} * \text{Cx a}} \quad \lor$$

$$x = \frac{-\text{ Cx b} - \sqrt{\text{Cx b pow 2} - \text{Cx 4} * \text{Cx a} * \text{Cx c}}}{\text{Cx 2} * \text{Cx a}}$$

Theorem 3.1 is a formally verified result for the two roots of the quadratic equation, known as the quadratic formula. Whereas, a, b and c are real coefficients of the given quadratic polynomial. The assumption A1 ensures that the given polynomial has a degree of 2.

The factorization theorem, i.e., Theorem 3.1, allows to formally reason about the stability condition in the formal definition of the stability, i.e., Definition 3.1. Next, we present the formal stability conditions on the two roots as

Lemma 1 *Complex Root Case*

$\vdash \forall$ a b c x.

A1 : a $\neq$ 0 $\land$

A2 : b pow 2 - 4 * a * c < 0 $\land$

A3 : 0 < $\frac{b}{a}$

$\Rightarrow$ stable (λ x. Cx a * x pow 2 + Cx b * x + Cx c)

Lemma 2 *Real Root Case 1*

$\vdash \forall$ a b c x.

A1 : a $\neq$ 0 $\land$

A2 : b pow 2 - 4 * a * c = 0 $\land$

A3 : 0 < $\frac{b}{a}$

$\Rightarrow$ stable (λ x. Cx a * x pow 2 + Cx b * x + Cx c)

Lemma 3 *Real Root Case 2*

⊢ ∀ a b c x.

  A1 : a < 0 ∧

  A2 : 0 < b pow 2 - 4 ∗ a ∗ c

  A3 : b < $\sqrt{\text{b pow 2} - 4 \ast a \ast c}$

  ⇒ stable (λ x. Cx a ∗ x pow 2 + Cx b ∗ x + Cx c)

Lemma 4 *Real Root Case 3*

⊢ ∀ a b c x.

  A1 : a < 0 ∧

  A2 : b pow 2 - 4 ∗ a ∗ c < 0 ∧

  A3 : $\sqrt{\text{b pow 2} - 4 \ast a \ast c}$ < - b

  ⇒ stable (λ x. Cx a ∗ x pow 2 + Cx b ∗ x + Cx c)

Lemma 5 *Real Root Case 4*

⊢ ∀ a b c x.

  A1 : 0 < a ∧

  A2 : 0 < b pow 2 - 4 ∗ a ∗ c ∧

  A3 : $\sqrt{\text{b pow 2} - 4 \ast a \ast c}$ < b

  ⇒ stable (λ x. Cx a ∗ x pow 2 + Cx b ∗ x + Cx c)

Lemma 6 *Real Root Case 5*

⊢ ∀ a b c x.

  A1 : 0 < a ∧

  A2 : 0 < b pow 2 - 4 ∗ a ∗ c ∧

  A3 : - b < $\sqrt{\text{b pow 2} - 4 \ast a \ast c}$

  ⇒ stable (λ x. Cx a ∗ x pow 2 + Cx b ∗ x + Cx c)

The formal verification of Lemmas 1–6 is conducted using complex, real and transcendental theories of the HOL Light theorem prover. The lemmas describe all the possible conditions in the terms of real coefficients of the stable quadratic polynomial. Figure 3.2 is a graphical illustration of the formal proofs of the lemmas for stable quadratic characteristic polynomials. Next, we utilize the above lemmas

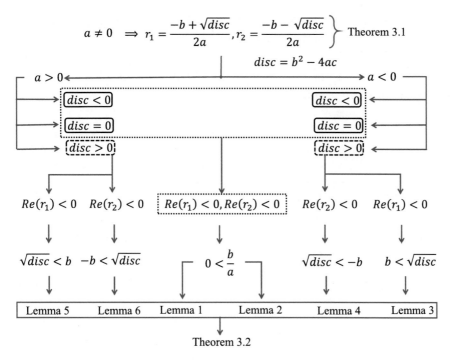

Fig. 3.2 Stability of quadratic polynomial

to formally verify the stability of quadratic polynomial, which covers all possible cases in the HOL Light theorem prover as

Theorem 3.2 $\vdash \forall$ a b c x.

A1 : $a \neq 0 \wedge$

A2 : $0 < \frac{b}{a} \wedge$ (b pow 2 - 4 * a * c < 0 $\vee$ b pow 2 - 4 * a * c = 0) $\vee$

0 < b pow 2 - 4 * a * c $\wedge$

(a < 0 $\wedge$ (b < $\sqrt{\text{b pow 2} - 4 * a * c}$ $\vee$

$\sqrt{\text{b pow 2} - 4 * a * c}$ < -b) $\vee$

(0 < a $\wedge$ ($\sqrt{\text{b pow 2} - 4 * a * c}$ < b $\vee$

-b < $\sqrt{\text{b pow 2} - 4 * a * c}$)

$\Rightarrow$ stable (λ x. Cx a * x pow 2 + Cx b * x + Cx c)

The above theorem exhaustively covers all the cases of the stable quadratic polynomials, which may arise due to the nature of the coefficients of the polynomial. In our proposed formalization, Theorem 3.2 plays a key role in formally verifying the stability of the cubic and the quartic polynomials.

3.4.2 Cubic Polynomial

In this section, we use the formally verified results of the quadratic polynomial and Definition 3.1 to formally verify the stability of cubic polynomial. This includes formal decomposition of cubic polynomial into its linear and quadratic factors and the formal stability analysis of the roots of the cubic roots.

We formally verify the factorization of cubic polynomial in HOL Light as follows:

Theorem 3.3 $\vdash \forall$ a b1 c1 d1 r x .

A1 : Cx b = Cx b1 + Cx a ∗ Cx r ∧

A2 : Cx c = Cx c1 + Cx b1 ∗ Cx r

A3 : Cx d = Cx c1 ∗ Cx r ∧

$\Rightarrow$ Cx a ∗ x pow 3 + Cx b ∗ xpow 2 + Cx c ∗ x + Cx d =

(x + Cx r) ∗ (Cx a ∗ x pow 2 + Cx b1 ∗ x + Cx c1)

The formally verified result provides an explicit relationship between the real coefficients of the cubic polynomial, i.e., a, b, c and d, linear factor, r and quadratic polynomial, i.e., a1, b1 and c1, as Assumption A1-A3.

Next, we present formally verified roots of the cubic polynomial using Definition 3.1, Lemmas 1–6 and Theorem 3.3 in HOL Light as:

Theorem 3.4 $\vdash \forall$ a b1 c1 d1 r x .

A1 : a $\neq$ 0 ∧

A2 : Cx b = Cx b1 + Cx a ∗ Cx r ∧

A3 : Cx c = Cx c1 + Cx b1 ∗ Cx r ∧

A4 : Cx d = Cx c1 ∗ Cx r

$\Rightarrow$ (Cx a ∗ x pow 3 + Cx b ∗ xpow 2 + Cx c ∗ x + Cx d = Cx 0)

$= ($ x = Cx r $\lor$ x $= \dfrac{-\,\mathrm{Cx\ b1} + \sqrt{\mathrm{Cx\ b1\ pow\ 2} - \mathrm{Cx\ 4} * \mathrm{Cx\ a} * \mathrm{Cx\ c1}}}{\mathrm{Cx\ 2} * \mathrm{Cx\ a}} \lor$

x $= \dfrac{-\,\mathrm{Cx\ b1} - \sqrt{\mathrm{Cx\ b1\ pow\ 2} - \mathrm{Cx\ 4} * \mathrm{Cx\ a} * \mathrm{Cx\ c1}}}{\mathrm{Cx\ 2} * \mathrm{Cx\ a}}$ $)$

The above theorem is formally verified using Theorem 3.3, which resulted in Assumptions A2–A4, whereas Assumption A1 ensures that the order of the polynomial is three. The formal verification of the linear root of the polynomial involved simple complex arithmetic, whereas the remaining two roots are formally verified using Theorem 3.1.

Finally, the above formalization is used to formally verify the stability of a cubic polynomial as

Theorem 3.5 $\vdash \forall$ a b1 c1 d1 r x.

A1 : a $\neq$ 0 $\wedge$ A2 : Cx b = Cx b1 + Cx a $*$ Cx r $\wedge$

A3 : Cx c = Cx c1 + Cx b1 $*$ Cx r $\wedge$ A4 : Cx d = Cx c1 $*$ Cx r

A5 : 0 < r $\vee$

((0 < $\frac{b1}{a}$ $\wedge$ (b1 pow 2 - 4 $*$ a $*$ c1 < 0 $\vee$

b1 pow 2 - 4 $*$ a $*$ c1 = 0))) $\vee$

(0 < b1 pow 2 - 4 $*$ a $*$ c1 $\wedge$

(a < 0 $\wedge$ (b1 $\sqrt{\text{b1 pow 2} - 4 * a * c1}$ $\vee$

$\sqrt{\text{b1 pow 2} - 4 * a * c1}$ < - b1) $\vee$

(0 < a $\wedge$ ($\sqrt{\text{b1 pow 2} - 4 * a * c1}$ < b1 $\vee$

- b < $\sqrt{\text{b1 pow 2} - 4 * a * c1}$)))

$\Rightarrow$ stable (λ x. Cx a $*$ x pow 3 + Cx b $*$ x pow 2 + Cx c $*$ x + Cx d)

The above theorem is formally verified using Lemmas 1–6 and the formal model of stability, i.e., Definition 3.1. Assumptions A1–A5 formally specify all the conditions on the real coefficients of a stable cubic polynomial.

3.4.3 Quartic Polynomial

In this section, we formally verify the stability of quartic polynomial using the formally verified results of quadratic polynomial, i.e., Theorem 3.2 and Definition 3.1. This includes formal factor decomposition into two quadratic factors: roots and stability of the quartic polynomial.

We formally verify the factorization of quartic polynomial in HOL Light as follows:

Theorem 3.6 $\vdash \forall$ a1 b1 c1 a2 b2 c2 x .

A1 : Cx a = Cx a1 $*$ Cx a2 $\wedge$

A2 : Cx b = Cx a1 $*$ Cx b2 + Cx a2 $*$ Cx b1 $\wedge$

A3 : Cx c = Cx a1 $*$ Cx c2 + Cx b1 $*$ Cx b2 + Cx a2 $*$ Cx c1 $\wedge$

A4 : Cx d = Cx b1 $*$ Cx c2 + Cx b2 $*$ Cx c1 $\wedge$

A5 : Cx e = Cx c1 $*$ Cx c2

$\Rightarrow$ (Cx a $*$ x pow 4 + Cx b $*$ x pow 3 + Cx c $*$ x pow 2 + Cx d $*$ x

$+$ Cx e = Cx 0) =

(((Cx a1 $*$ x pow 2 + Cx b1 $*$ x + Cx c1) $*$

(Cx a2 $*$ x pow 2 + Cx b2 $*$ x + Cx c2))

The formally verified result explicitly specifies the relationship among the real coefficients of the quartic polynomial and quadratic factors, i.e., Assumptions A1-A5.

We now use the above formal results to formally verify the roots of the quartic polynomial in HOL Light as:

Theorem 3.7 $\vdash \forall$ a1 b1 c1 a2 b2 c2 x .

A1 : a $\neq$ 0 $\wedge$ A2 : Cx a = Cx a1 $*$ Cx a2 $\wedge$

A3 : Cx b = Cx a1 $*$ Cx b2 + Cx a2 $*$ Cx b1 $\wedge$

A4 : Cx c = Cx a1 $*$ Cx c2 + Cx b1 $*$ Cx b2 + Cx a2 $*$ Cx c1 $\wedge$

A5 : Cx d = Cx b1 $*$ Cx c2 + Cx b2 $*$ Cx c1 $\wedge$

A6 : Cx e = Cx c1 $*$ Cx c2

$\Rightarrow$ (Cx a $*$ x pow 4 + Cx b $*$ x pow 3 + Cx c $*$ x pow 2 + Cx d $*$ x

$+$ Cx e = Cx 0) =

$$(\ x = \frac{-\,Cx\ b1 + \sqrt{Cx\ b1\ pow\ 2 - Cx\ 4 * Cx\ a1 * Cx\ c1}}{Cx\ 2 * Cx\ a1} \ \vee$$

$$x = \frac{-\,Cx\ b1 - \sqrt{Cx\ b1\ pow\ 2 - Cx\ 4 * Cx\ a1 * Cx\ c1}}{Cx\ 2 * Cx\ a1} \ \vee$$

$$x = \frac{-\,Cx\ b2 + \sqrt{Cx\ b2\ pow\ 2 - Cx\ 4 * Cx\ a2 * Cx\ c2}}{Cx\ 2 * Cx\ a2} \ \vee$$

$$x = \frac{-\,Cx\ b2 - \sqrt{Cx\ b2\ pow\ 2 - Cx\ 4 * Cx\ a2 * Cx\ c2}}{Cx\ 2 * Cx\ a2} \ )$$

Based on the Assumptions A1–A6, Theorem 3.7 formally verifies the roots of the quartic polynomial. Whereas, Theorems 3.1 and 3.6 play a major role in the formal verification of the roots of the quartic polynomial.

The above formalization allows to formally reason and verify the stability of quartic polynomial in HOL Light as:

Theorem 3.8 $\vdash \forall$ a1 b1 c1 a2 b2 c2 x .

A1 : a $\neq$ 0 $\wedge$ A2 : Cx a = Cx a1 $*$ Cx a2 $\wedge$

A3 : Cx b = Cx a1 $*$ Cx b2 + Cx a2 $*$ Cx b1 $\wedge$

A4 : Cx c = Cx a1 $*$ Cx c2 + Cx b1 $*$ Cx b2 + Cx a2 $*$ Cx c1 $\wedge$

A5 : Cx d = Cx b1 $*$ Cx c2 + Cx b2 $*$ Cx c1 $\wedge$

A6 : Cx e = Cx c1 $*$ Cx c2 $\wedge$

A7 : (0 < $\frac{b1}{a1}$ $\wedge$ (b1 pow 2 - 4 $*$ a1$*$ c1 < 0 $\vee$

b1 pow 2 - 4 $*$ a1$*$ c1 = 0)) $\vee$

(b1 pow 2 - 4 $*$ a1$*$ c1 < 0 $\wedge$

(a1 < 0 $\wedge$ (b1 < $\sqrt{\text{b1 pow 2} - 4 * a1 * c1}$ $\vee$

$\sqrt{\text{b1 pow 2} - 4 * a1 * c1}$ < - b1) $\vee$

(0 < a1 $\wedge$ ($\sqrt{\text{b1 pow 2} - 4 * a1 * c1}$ < b1 $\vee$

- b1 < $\sqrt{\text{b1 pow 2} - 4 * a1 * c1}$)) $\vee$

(0 < $\frac{b2}{a2}$ $\wedge$ (0 < b2 pow 2 - 4 $*$ a2$*$ c2 $\vee$

b2 pow 2 - 4 $*$ a2$*$ c2 = 0)) $\vee$

(b2 pow 2 - 4 $*$ a2$*$ c2 < 0 $\wedge$

(a2 < 0 $\wedge$ (b2 < $\sqrt{\text{b2 pow 2} - 4 * a2 * c2}$ $\vee$

$\sqrt{\text{b2 pow 2} - 4 * a2 * c2}$ < - b1) $\vee$

(0 < a2 $\wedge$ ($\sqrt{\text{b2 pow 2} - 4 * a2 * c2}$ < b2 $\vee$

- b2 < $\sqrt{\text{b2 pow 2} - 4 * a2 * c2}$))

$\Rightarrow$ stable (λ x. (Cx a $*$ x pow 4 + Cx b $*$ x pow 3 + Cx c $*$ x pow 2

+ Cx d $*$ x + Cx e)

Theorem 3.8 is a formally verified result of the stability of quartic polynomial based on the Assumptions A1–A7. Theorems 3.2 and 3.7 are mainly used to formally verify the stability of the quartic polynomial.

The proposed methodology resulted in obtaining an exhaustive set of assumptions that are required for the factorization and stability theorems of the polynomials upto the fourth order. These assumptions explicitly states the conditions on the real coefficients of the polynomials. The real coefficients of the polynomial correspond to the important design parameters of the underlying power converter systems, and hence can help design highly reliable and efficient power converters. Notably, the proposed formalization is generic and can be readily used to formally verify many engineering control systems which are represented using transfer functions in frequency domain. The proof script of the proposed formalization is available online [9], and has 5000 lines of HOL Light code which took about 380 man hours of development time.

3.5 Application: Power Converter Controllers Used in Smart Grids

Intermittent energy flow from renewable energy sources, such as wind turbines, are smoothed using efficient designing of controllers for power converters, as shown in Fig. 3.3. We formally verify the stability of an H^∞ current, H^∞ voltage and H^∞ repetitive current controllers designed for the power converters to enhance the efficiency of smart grids [10], using our proposed formalization. H^∞ [11] and repetitive control [12] methods are used to process the intermittent energy flow from renewable energy sources to achieve reliable and secure smart grid operations.

The transfer function of an H^∞ current controller is reported in [10] as

$$[TF]_i = \frac{1.7998 * 10^9 (s + 300)}{s^2 + 4.33403 * 10^8 s + 1.10517 * 10^{12}} \tag{3.1}$$

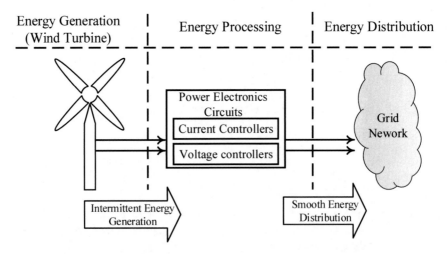

Fig. 3.3 Efficient energy harvesting using Power converter controllers in smart grids

We use the denominator of (3.1), which is a characteristic equation of the H^∞ current controller, to formally verify the stability of the controller using the stability formalization as

Theorem 3.9 $\vdash \forall$ a b c s .

```
stable(λx.Cx1 * s pow 2 + Cx 4.3340 * 10⁸ * x + Cx 1.10517 * 10¹²)
```

The above theorem is formally verified using Theorem 3.2 and formal complex arithmetic reasoning.

Similarly, the transfer function for an H^∞ voltage controller is reported in [10] as

$$[TF]_v = \frac{748.649(s^2 + 6954\,s + 3.026 * 10^8)}{s^3 + 10519\,s^2 + 3.246 * 10^8\,s + 7.7596 * 10^7} \tag{3.2}$$

The denominator of the (3.2) is a polynomial of third order and is formally verified using Theorem 3.5 in HOL Light theorem prover as:

Theorem 3.10 $\vdash \forall$ a b1 c1 d1 r s .

A1 : $a = 1 \wedge$ A2 : $b1 = 79669 \wedge$ A3 : $c1 = 3.043 * 10^8 \wedge$ A4 : $r = 2550$

$\Rightarrow$ stable (λx. Cx1 * s pow 3 + Cx10519 * s pow 2 + Cx3.246 * 10⁸ * s

+ Cx7.7596 * 10⁷)

Next, we formally verify the stability of a H^∞ repetitive current controller [10], which has the following transfer function:

$$[TF]_{vr} = \frac{8.63 * 10^8(s + 10^4)(s + 1000)(s + 80)}{s^4 + 1.55 * 10^8\,s^3 + 1.83 * 10^{13}s^2 + 1.43 * 10^{17}s + 1.08 * 10^{19}} \tag{3.3}$$

The fourth order characteristic equation of (3.3) is formally verified using Theorem 3.8 as:

Theorem 3.11 $\vdash \forall$ a1 b1 c1 a2 b2 c2 s .

A1 : $a1 = 1 \wedge$ A2 : $b1 = 1.557 * 10^7 \wedge$ A3 : $c1 = 1.70538 * 10^3 \wedge$

A4 : $a2 = 1 \wedge$ A5 : $b2 = 8.403 * 10^3 \wedge$ A6 : $c2 = 6.375 * 10^5$

$\Rightarrow$ stable (λx. Cx1 * s pow 4 + Cx1.55 * 10⁸ * s pow 3 +

Cx 1.83 * 10¹³ * s pow 2 + 1.43 * 10¹⁷ * s + Cx1.08 * 10¹⁹)

We utilize the proposed formalization, i.e., Theorems 3.2, 3.5 and 3.8, to formally verify the stability of smart grid controllers for renewable energy sources, i.e., Theorems 3.9–3.11. Due to generic nature of the proposed formalization, the stability of the power converters for smart grids were formally verified with only few lines of code mainly using formal real and complex arithmetic. The primary advantage of the formal verification is the exhaustive set of assumptions obtained using the theorem proving based methodology. The assumptions in Theorems 3.9–3.11 are available in terms of the coefficients of the transfer functions. Therefore, explicit conditions on these parameters can be very useful in the design of safe, reliable and secure controllers for smart grids.

3.6   Summary

This chapter presented the formalization of the stability of the control systems. The formalization consisted of the formal modeling of the stability criterion and formal verification of the stability of the characteristic equations upto the fourth order. The notable formal results included formally verified theorems for second, third and fourth order characteristic equations which cover a wide spectrum of control systems for many interesting engineering applications. To illustrate the usefulness of the proposed formalizations, we conducted the formal stability analysis of the H^∞ current, voltage and repetitive controllers for the smart grids. The exhaustive and sound stability analysis of the power converters using theorem proving ensures the secure and reliable generation and distribution of energy in the smart grid networks.

References

1. J. Harrison, HOL-Light Theorem Prover. https://www.cl.cam.ac.uk/~jrh13/hol-light/. Accessed 1 Mar 2021
2. S.H. Taqdees, O. Hasan, Formalization of Laplace transform using the multivariable calculus theory of HOL-light, in *International Conference on Logic for Programming Artificial Intelligence and Reasoning* (Springer, Berlin, 2013), pp. 744–758
3. O. Hasan, M. Ahmad, Formal analysis of steady state errors in feedback control systems using hol-light, in *Proceedings of the Conference on Design, Automation and Test in Europe* (EDA Consortium, 2013), pp. 1423–1426
4. M. Ahmad, O. Hasan, Formal verification of steady-state errors in unity-feedback control systems, in *International Workshop on Formal Methods for Industrial Critical Systems* (Springer, Berlin, 2014), pp. 1–15
5. U. Siddique, V. Aravantinos, S. Tahar, Formal stability analysis of optical resonators, in *NASA Formal Methods Symposium*, LNCS, vol. 7871 (Springer, Berlin, 2013), pp. 368–382
6. A. Rashid, U. Siddique, O. Hasan, Formal verification of platoon control strategies, in *International Conference on Software Engineering and Formal Methods* (Springer, Berlin, 2018), pp. 223–238
7. A. Rashid, O. Hasan, Formal analysis of linear control systems using theorem proving, in *International Conference on Formal Engineering Methods* (Springer, Berlin, 2017), pp. 345–361
8. M.U. Sanwal, O. Hasan, Formally analyzing continuous aspects of cyber-physical systems modeled by homogeneous linear differential equations, in *International Workshop on Design, Modeling, and Evaluation of Cyber Physical Systems*, LNCS, vol. 9361 (Springer, Berlin, 2015), pp. 132–146

9. A. Ahmed, Formal Stability Analysis of Control Systems. http://save.seecs.nust.edu.pk/projects/fsacs/. Accessed Mar 2021
10. Q.C. Zhong, T. Hornik, *Control of Power Inverters in Renewable Energy and Smart Grid Integration*, vol. 97 (Wiley, New York, 2012)
11. A.A. Stoorvogel, *The* H^∞ *Control Problem: A State Space Approach* (Citeseer, 1992)
12. T. Hornik, Q.C. Zhong, A current-control strategy for voltage-source inverters in microgrids based on H^∞ and repetitive control. IEEE Trans. Power Electron **26**(3), 943–952 (2011)

Chapter 4
Formalization of Cost and Utility in Microeconomics

In smart grids, cost and utility modeling is used to assess the cost of energy generation from energy sources such as thermal plants and design incentives in DR programs, respectively. Cost and utility modeling, mainly, uses convex and differential theories to model the behaviors of grid actors, such as consumers. Traditionally, paper-and-pencil proof methods and computer-based tools are used to investigate the mathematical properties of cost and utility models. However, these techniques do not provide an accurate analysis due to their inability to exhaustively specify and verify the mathematical properties of the cost and utility models. Whereas accurate analysis is direly needed in mission-critical applications of smart grids such as energy generation and DR programs. To overcome the issues pertaining to the above-mentioned techniques, in this chapter, we present a theorem proving based logical framework to formally analyze and specify the mathematical properties of cost and utility modeling. The logical framework is used to formally verify the estimates of coefficients of cost function for a thermal power plant.

4.1 Introduction

In this chapter, we present a logical framework for the behavioral modeling in microeconomics. This logical framework enables formally specifying and reasoning about the cost, utility and first-order condition models within the sound core of the HOL-light theorem prover. A formal analysis of these models in higher-order logic results in the certified and mechanized formal proofs of given models. As a case study, we provide the formal modeling and analysis of the behavioral modeling in the electricity market based upon the polynomial functions, up to the fourth order.

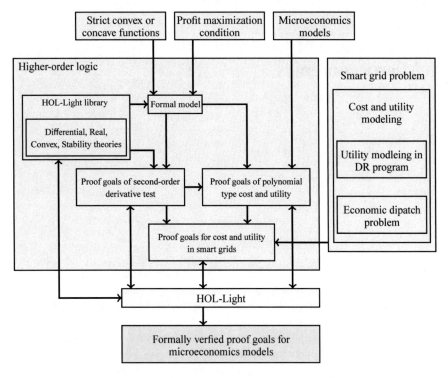

Fig. 4.1 Theorem proving based methodology for microeconomics modeling

The formally verified results, for the behavioral modeling of electricity market, provide necessary conditions on the functions to accurately model the cost, utility and first-order condition. Finally, we utilize this behavioral modeling to formally verify the applications of cost, utility and profit-maximization models in economic dispatch and dynamic-pricing problems in the electricity market (as shown in Fig. 4.1).

The rest of the chapter is organized as follows: Section 4.2 presents the proposed methodology and, based on this, Sects. 4.3–4.4 provides the formalization of second-order derivative test for strict convexity or concavity, cost, utility and first-order condition models in higher-order logic theorem prover. Section 4.5 utilizes the basic logical framework to formally specify and verify the mathematical behavioral modeling in DR programs for a polynomial type of modeling functions. Finally, Sect. 4.6 concludes the paper.

4.2 Proposed Methodology

In this section, we present the proposed methodology, shown in Fig. (4.1), for conducting the formal behavioral modeling using the microeconomics theory:

1. First, we formally model the strict concavity or convexity and first-order condition (1.3) notions in higher-order logic and formally verify the second-order derivative test (1.5a) for strict concave or convex functions. This allows to formally specify and reason about the consumer or firm behavior in the sound core of the HOL Light theorem prover.
2. Next, we provide formally verified results of the cost and utility models and profit-maximization conditions of polynomial type of functions upto the fourth order, mainly using the formalization from Step 1 and formal factorization theorems for the polynomials (given in Chap. 3) .
3. Finally, the above logical framework is used to formally verify the polynomial type of cost and utility functions used in the electricity dispatch and DR economic problems in smart grids.

4.3 Formalization of Microeconomics Concepts

In this section, we present a higher-order logic formalization of the strict convex or concave functions and second-order derivative test for the strict convexity or concavity of a real function. These formally verified results are further employed to formally model the cost and utility functions in microeconomics. We also present formal modeling of the first-order condition in the HOL Light theorem prover.

4.3.1 Formalization of Strict Convexity and Concavity

We formally model the strict convex function in higher-order logic as:

Definition 4.1 $\vdash \forall$ s f. f real_strict_convex_on s =
$(\forall$ x y u v. x $\in$ s $\land$ y $\in$ s $\land$ &0 $<$ u $\land$ &0 $<$ v $\land$ u + v $<$ &1
$\Rightarrow$ f (u * x + v * y) $<$ u * f x + v * f y)

In Definition 4.1, real_strict_convex_on is a higher-order logic function which accepts a real function, $f : R \rightarrow R$, and the real interval, $s : R \rightarrow bool$. The formal definition of the strict convex function differs from Definition 2.11 by a strict inequality.

Next, we formally verify the mathematical properties of scalar multiplication and addition of the strict convex functions in the HOL Light theorem prover.

Lemma 1 *constant multiplier property*
⊢ ∀ f s c.

A1 : 0 < c ∧ A2 : f real_strict_convex_on s

$$\Rightarrow ((\lambda x. c * f x) \text{ real\_strict\_convex\_on } s).$$

Lemma 1 formally verifies that the multiplication of a strictly positive scalar number, i.e., $0 < c$, with the given strict convex function does not affect the strict convexity of the function.

Similarly, the addition property of the two strict convex functions resulting in another strictly convex function is formally verified in the HOL Light theorem prover as

Lemma 2 *Addition property*
⊢ ∀ f s g.

A1 : f real_strict_convex_on s   A2 : g real_strict_convex_on s

$$\Rightarrow (\lambda x. f x + g x \text{ real\_strict\_convex\_on } s)$$

The above theorem formally verifies that the sum of the two strictly convex functions f and g results in a strict convex function, i.e., f + g.

Next, we formally verify the second-order test which is used to assess the strict convexity of the given function, in HOL Light as

Theorem 4.1 ⊢ ∀ f f' f'' s .
A1 : ¬(∃a. s = { a }) ∧
A2 : (∀x. x ∈ s ⇒ (f has_vector_derivative f' x) (atreal x within x))
A3 : (∀x. x ∈ s ⇒ (f' has_vector_derivative f'' x) (atreal x within x))
A4 : f real_strict_convex_on s
$$\Rightarrow (0 < f'')$$

The above theorem requires a function, $f : (R \rightarrow R)$, its first and second-order derivative, i.e., f' and f", and real interval, s, for which second-order derivative test is required to be formally verified. The Assumption A1 avoids single tone real interval so that the function can be analyzed for differentiability. Assumptions A2 and A3 ensures that the function has well-defined first and second-order derivatives within the given interval, respectively. Assumption A4 ensures that the function is strictly convex. Based on these assumptions, the second-order derivative of the function is formally verified to be strictly positive. The verification of Theorem 4.1 involves mean value theorem (Theorem 2.5) and basic definition of strict convexity (Definition 4.1).

As strict convex and concave functions are mathematically related, i.e., the negative of the strict convex function is a strict concave function and vice versa. Thus, we utilize this mathematical relationship to formally verify the second-order derivative test of the strict concave functions in the HOL Light theorem prover as

Theorem 4.2 $\vdash \forall$ f f' f'' s .
A1 : $\neg(\exists a.\ s = \{\ a\ \}) \wedge$
A2 : $(\forall x.\ x \in s \Rightarrow (\ f\ \text{has\_vector\_derivative}\ f'\ x\ )(\ \text{atreal}\ x\ \text{within}\ x\ ))$
A3 : $(\forall x.\ x \in s \Rightarrow (\ f'\ \text{has\_vector\_derivative}\ f''\ x\ )(\ \text{atreal}\ x\ \text{within}\ x\ ))$
A4 : $-$f real_strict_convex_on s
$\Rightarrow (\ f'' < 0\ )$

Assumption A4 formally specifies the strict concavity of the given function by using
the negative operation on the given function. This results in the strictly negative
second-order derivative of the function in the conclusion of the above theorem.
Theorem 4.2 is formally verified following the same line of action as was used
for Theorem 4.1. The above formalizations allow to formally model the cost and
utility functions by stating the mathematical properties of continuity, differentiability,
monotonicity and convexity or concavity in the HOL Light theorem prover.

Mathematical properties of the cost function are formally modeled in HOL Light
as

Definition 4.2 $\vdash \forall$ f x.convex_cost_func f s x =
(f real_continuous_on s) $\wedge$ (f real_differentiable_on s) $\wedge$
($\forall$ x y. (x $\in$ s) $\wedge$ (y $\in$ s) $\wedge (x \leq y) \Rightarrow (f x \leq f y)) \wedge$
(f real_convex_on s))

Definition 4.3 $\vdash \forall$ f x. strict_convex_cost_func f s x =
(f real_continuous_on s) $\wedge$ (f real_differentiable_on s) $\wedge$
($\forall$ x y. (x $\in$ s) $\wedge$ (y $\in$ s) (x $<$ y) $\Rightarrow$ (f x $<$ f y) $\wedge$
(f real_strict_convex_on s))

Similarly, mathematical properties of the utility function are formally modeled in
HOL Light as

Definition 4.4 $\vdash \forall$ f x. concave_utility_func f s x =
(f real_continuous_on s) $\wedge$ (f real_differentiable_on s) $\wedge$
($\forall$ x y. (x $\in$ s) $\wedge$ (y $\in$ s) $\wedge$ (f x $\leq$ f y)) $\wedge$
($-$f real_convex_on s)

Definition 4.5 $\vdash \forall$ f x. strict_concave_utility_func f s x =
(f real_continuous_on s) $\wedge$ (f real_differentiable_on s) $\wedge$
($\forall$ x y. (x $\in$ s) $\wedge$ (y $\in$ s) $\wedge$ (f x $<$ f y)) $\wedge$
($-$f real_strict_convex_on s)

To conduct the formal profit-maximization analysis, we model the first-order
condition (1.3) in higher-order logic as

Definition 4.6 $\vdash \forall$ f x. first_order_cond f x =(real_derivative f x
= &0)

Definition 4.6 formally models the first-order condition using the higher-order-logic definition of the derivative of a real function, i.e., `real_derivative`. It accepts a real function, `f`, and the evaluation point, `x`, to formally specify the first-order condition.

The formalization developed in this section provides a logical framework that enables the formal analysis and verification of the cost, utility and profit-maximization concepts of microeconomics theory.

4.4 Case Study: Formal Behavioral Modeling Based on Polynomial Functions

Polynomial functions are widely used in the microeconomics analysis to model the behaviors of the consumers and firms [1–3]. In this section, we use our proposed methodology to formalize cost and utility modeling using polynomial functions. We also provide a formalization of the first-order condition of a profit function which uses the polynomial type of the cost function.

4.4.1 Polynomial Type of Cost Functions

A quadratic polynomial as a cost function is formally verified in the HOL Light theorem prover as

Theorem 4.3 $\vdash \forall$ x s a b c.
A1 : (is_realinterval s) $\wedge$ A2 : (real_open s) $\wedge$ A3 : $\neg(\exists$a1. s = { a1 }) $\wedge$
A4 : strict_convex_cost_func (λx. Cx a $*$ x pow 2 + Cx b $*$ x + Cx c) $\wedge$
A5 : (x $\in$ s) $\wedge$ A6 : ($-\frac{b}{2a}$ < x)   A7 : (0 < a) $\wedge$
$\Rightarrow$ 0 < real_derivative (λx. Cx a $*$ x pow 2 + Cx b $*$ x + Cx c) $\wedge$
0 < real_derivative (real_derivative (λx. Cx a $*$ x pow 2 +
Cx b $*$ x + Cx c))

In Theorem 4.3, Assumptions A1- A3 ensure that the given interval, s, is real, open and non-singular. Assumption A4 ensures that the given quadratic polynomial is strictly convex. Whereas Assumptions A5- A7 impose some mandatory conditions on the variable x and coefficient a for the validity of the theorem. Finally, the conclusion of the theorem ensures that the first and second-order derivatives of the quadratic polynomial are strictly positive.

Next, we formally verify the cubic polynomial as a cost function in HOL Light theorem prover as

Theorem 4.4 $\vdash \forall$ x s a b c d.
A1 : (is_realinterval s) $\wedge$ A2 : (real_open s) $\wedge$ A3 : $\neg(\exists$a1. s = { a1 })
A4 : strict_convex_cost_func (λx. Cx a $*$ x pow 3 + Cx b $*$ xpow 2 + Cx c $*$ x
+ Cx d)
A5 : (x $\in$ s) $\wedge$
A6 : ($-\frac{b}{3a}$ < x)) $\wedge$ A7 : ($\frac{-b-\sqrt{b^2+3*a*c}}{3a}$ < x) $\vee$ ($\frac{-b+\sqrt{b^2+3*a*c}}{3a}$ < x)

$\Rightarrow 0 <$ real_derivative $(\lambda x.$ Cx a $*$ x pow $3 +$ Cx b $*$ xpow $2 +$ Cx c $*$ x $+$ Cx d
Cx c $*$ x $+$ Cx d $)) \wedge$
$0 <$ real_derivative (real_derivative $(\lambda x.$ Cx a $*$ x pow $3 +$ Cx b $*$ xpow 2
$+$
Cx c $*$ x $+$ Cx d $)))$

Based on Assumptions A1–A7, Theorem 4.4 formally verifies the cubic polynomial as a cost function. In particular, Assumptions A6–A7 formally specify the domain variable of the cubic polynomial for which the polynomial satisfies the mathematical properties of the cost function of an economic agent.

Theorem 4.5 $\vdash \forall$ x s a b c d e.
A1 : (is_realinterval s) $\wedge$ A2 : (real_open s) $\wedge$ A3 : $\neg(\exists$a1. s $= \{$ a1 $\})$
A4 : strict_convex_cost_func $(\lambda x.$ Cx a $*$ xpow $4 +$ Cxb $*$ xpow $3 +$
Cx c $*$ xpow $2 +$ Cxd $*$ x $+$ Cxe $)$
A5 : (x $\in$ s)$\wedge$ A6 : ($a_1 = 4 * a$) $\wedge$
A7 : ($b_1 + a_1 * r = 3 * b$) $\wedge$ A8 : ($c_1 + b_1 * r = 2 * c$) $\wedge$ A9 : ($- r * c_1 = d$) $\wedge$
A10 : ($-r \le x$ $\vee$ $\frac{b_1 - \sqrt{b_1^2 - 4*a_1*c_1}}{2a_1} \le x$ $\vee$ $\frac{b_1 + \sqrt{b_1^2 - 4*a_1*c_1}}{2a_1} \le x$) $\wedge$
A11 : ($\frac{3*b - \sqrt{9*b^2 + 24*a*c}}{12a} < x \vee$ $\frac{3*b + \sqrt{9*b^2 + 24*a*c}}{12a} < x$)
$\Rightarrow 0 <$ real_derivative $(\lambda x.$ Cx a $*$ xpow $4 +$ Cxb $*$ xpow $3 +$ Cx c $*$ xpow 2
$+$ Cxd $*$ x $+$ Cxe $)) \wedge$
$0 <$ real_derivative (real_derivative $(\lambda x.$ Cxa $*$ xpow $4 +$ Cxb $*$ xpow $3 +$
Cx c $*$ xpow $2 +$ Cxd $*$ x $+$ Cxe $))$

Based on the Assumptions A1–A11, Theorem 4.5 formally verifies the quartic polynomial modeling as a cost function. Assumptions A6–A9 formally specify the relationship of the coefficients of the quartic polynomial and the corresponding quadratic factors. Assumptions A10–A11 formally specify the domain for modeling the quartic polynomial as a cost function.

Theorems 4.3–4.5 formally verify the quadratic, cubic and quartic polynomials as cost functions. The formal verification mainly involved the use of the Definition 4.3, Lemmas 1, 2, Theorem 4.1 and stability formalization from Chap. 3 (Theorems 3.3, 3.4, 3.7).

4.4.2 Polynomial Type of Utility Function

In this section, we formally verify the polynomial type of the utility functions in the HOL Light theorem prover.

A quadratic type of utility function is formally verified as

Theorem 4.6 $\vdash \forall$ f x s a.
A1 : (is_realinterval s) $\wedge$ A2 : (real_open s) $\wedge$
A3 : $\neg(\exists$a. s $= \{$ a $\})$ $\wedge$ A4 : (x $\in$ s) $\wedge$
A5 : ($\forall$ x. concave_utility_func $(\lambda x.$ (Cx a $*$ x pow $2 +$ Cx b $*$ x $+$ Cx c)) $\wedge$
A6 : ($-\frac{b}{2a} \le x$)$\wedge$ A7 : (a < 0)
$\Rightarrow$ $0 \le$ real_derivative $(\lambda x.$ Cx a $*$ x pow $2 +$ Cx b $*$ x $+$ Cx c) $\wedge$
real_derivative $(\lambda$ x. real_derivative $(\lambda$ x. Cx a $*$ x pow $2 +$
c Cx b $*$ x $+$ Cx c $)) \le 0$

In Theorem 4.6, Assumptions A1–A4 formally specifies that the domain variable, x, is in the real, open and non-singular interval, s. Assumption A5 formally specifies the utility conditions on the quadratic polynomial, using Definition 4.5, whereas Assumptions A6–A7 formally specify the range of the domain variable, $-\frac{b}{2a} \le$ x, and real coefficient a, i.e., a < 0. Based on these assumptions, Theorem 4.6 formally verifies the decreasing first and second-order derivatives of the given quadratic type of the cost function.

Theorem 4.7 $\vdash \forall$ f x s a b c d.
A1 : (is_realinterval s) $\wedge$ A2 : (real_open s) $\wedge$
A3 : $\neg(\exists a1.$ s $= \{$ a1 $\}) \wedge$ A4 : (x $\in$ s) $\wedge$
A5 : ($\forall$ x. concave_utility_func (λx. (Cx a $*$ x pow 3 $+$ Cx b
$*$ xpow 2 $+$ Cx c $*$ x $+$ Cx d) $\wedge$
A6 : ($\frac{b-\sqrt{b^2+3*a*c}}{3a} \le$ x) $\vee$ ($\frac{b+\sqrt{b^2+3*a*c}}{3a} \le$ x) **A7:** (x $\le -\frac{b}{3a}$)
$\Rightarrow$ (0 $\le$ real_derivative (λx. (Cx a $*$ x pow 3 $+$ Cx b $*$ xpow 2 $+$
Cx c $*$ x $+$ Cx d)) $\wedge$
real_derivative (λx. real_derivative (λx. (Cx a $*$ x pow 3 $+$
Cx b $*$ xpow 2 $+$ Cx c $*$ x $+$ Cx d) $\le$ 0)

The above theorem formally verifies the utility function modeled using cubic polynomial.

Finally, we formally verify the quartic polynomial as a utility function in higher-order logic as

Theorem 4.8 $\vdash \forall$ f x s a b c d.
A1 : (is_realinterval s) $\wedge$ A2 : (real_open s) $\wedge$ A3 : $\neg(\exists a.$ s $= \{$ a $\}) \wedge$
A4 : (x $\in$ s) $\wedge$ A5 : ($\forall$ x. concave_utility_func (λx. (Cx a $*$ xpow 4 $+$
Cx b $*$ xpow 3 $+$ Cx c $*$ xpow 2 $+$ Cx d $*$ x $+$ Cx e)) $\wedge$ A6 : ($a_1 = 4 * a$) $\wedge$
A7 : ($b_1 + a_1 * r = 3 * b$) $\wedge$ A8 : ($c_1 + b_1 * r = 2 * c$) $\wedge$ A9 : ($- r * c_1 = d$) $\wedge$
A10 : ($-r \le$ x $\vee$ $\frac{b_1-\sqrt{b_1^2-4*a_1*c_1}}{2a_1} \le$ x $\vee$ $\frac{b_1+\sqrt{b_1^2-4*a_1*c_1}}{2a_1} \le$ x) $\wedge$
A11 : (x $\le \frac{3*b-\sqrt{9*b^2+24*a*c}}{12a}$ $\vee$ x $\le \frac{3*b+\sqrt{9*b^2+24*a*c}}{12a}$)
$\Rightarrow$ (0 $\le$ real_derivative λx. (Cx a $*$ xpow 4 $+$ Cx b $*$ xpow 3 $+$
Cx c $*$ xpow 2 $+$ Cx d $*$ x $+$ Cx e)) $\wedge$
real_derivative (λx. real_derivative (λx. (Cx a $*$ xpow 4 $+$ Cx b $*$ xpow 3 $+$
Cx c $*$ xpow 2 $+$ Cx d $*$ x $+$ Cx e)) $\le$ 0

Assumptions A6–A9 formally specify the relationship among the coefficients of the quartic polynomial and its corresponding quadratic factors, whereas Assumptions A10–A11, formally specify the range of the domain variable, x, in terms of the coefficients of factors of the quartic polynomial. Based on these assumptions, the decreasing first and second-order derivatives of the quartic polynomial are formally verified.

The formal utility modeling using quadratic, cubic and quartic polynomials, i.e., Theorems 4.6–4.8, mainly rely upon Definition 4.4 and Theorems 3.1–3.8. Definition 4.4 allows to specify the differentiability, continuity and concavity properties, whereas stability formalizations allow to formally specify the domain in terms of the polynomial coefficients.

4.4.3 First-Order Condition

In this section, we formally verify the first-order condition for profit functions of the mathematical form

$$profit(x) = rate * x - C(x)$$

In the above profit equation, $rate * x$ is a linear revenue function, where $rate$ is the rate of the electricity for the given amount of the electricity consumption, i.e., x. We formally verify profit functions that are designed using the polynomial type of the cost functions ($C(x)$).

In this regard, we present a formally verified result for a profit function based on the quadratic type of the cost function in the HOL Light theorem prover as

Theorem 4.9 $\vdash \forall$ a b c x .
A1 : a $\neq$ 0 $\wedge$
A2 : x = $\frac{r-b}{2a}$
$\Rightarrow$ first_order_cond (λ x. r $*$ x $-$ (a $*$ x pow 2 $+$ b $*$ x $+$ c)) x

In Theorem 4.9, Assumption A1 ensures that the degree of the given polynomial is 2. Assumption A2 formally specifies the critical or stationary point for the first-order derivative of the quadratic function in terms of the electricity rate, r, and polynomial coefficients.

The formal verification of Theorem 4.9 involved taking derivative of the given profit function which reduced the profit function into a first degree polynomial in x. The Assumption A2 then allows to formally verify the first-order condition for profit function modeled using the quadratic cost function.

Similarly, the profit function for the cubic cost function is formally verified as

Theorem 4.10 $\vdash \forall$ a b c d x .
A1 : a $\neq$ 0 $\wedge$
A2 : x = $\frac{- \text{Cx b} + \sqrt{\text{Cx b pow 2} - \text{Cx 4} * \text{Cx a} * \text{Cx c}}}{\text{Cx 2} * \text{Cx a}}$ $\vee$
x = $\frac{- \text{Cx b} - \sqrt{\text{Cx b pow 2} - \text{Cx 4} * \text{Cx a} * \text{Cx c}}}{\text{Cx 2} * \text{Cx a}}$
A3 : 0 < b pow 2 - 3 $*$ a $*$ (c - r) $\wedge$
(a < 0 $\wedge$ (b < $\sqrt{\text{b pow 2} - 3 * a * (c - r)}$ $\vee$
$\sqrt{\text{b pow 2} - 3 * a * (c - r)}$ < - b) $\vee$
(0 < a $\wedge$ ($\sqrt{\text{b pow 2} - 3 * a * (c - r)}$ < b $\vee$
- b < $\sqrt{\text{b pow 2} - 3 * a * (c - r)}$)
$\Rightarrow$ first_order_cond (λ x. r $*$ x $-$ (a $*$ x pow 3 $+$
b $*$ x pow 2 $+$ c $*$ x $+$ d)) x

In Theorem 4.10, Assumption A2 arises from the roots of the quadratic polynomial and Assumption A3 ensures that the roots of the polynomial are real.

The formal verification involved taking derivative of the profit function and using Theorem 3.1 and Assumption A2 to formally verify the profit maximization of the given profit function at critical points. Whereas, Assumption A3 is the consequence of the considered real roots of the quadratic equation that reduce to the inequality conditions on the coefficients of the quadratic type of the polynomial.

Finally, we present the formal verification of the profit function based on the quartic type of polynomial cost, in HOL Light theorem prover as

Theorem 4.11 $\vdash \forall$ a b c d e x .

A1 : $a \neq 0 \wedge$ A2 : $a1 = 4a \wedge$

A3 : $b1 + a1 * r = 3b \wedge$ A4 : $c1 + b1 * r = 2c$

A5 : $c1 + b1 * r = 2c$ A6 : $d - r = c1 * r$

A7 : ($x = Cx\ r \vee x = \dfrac{-Cx\ b1 + \sqrt{Cx\ b1\ pow\ 2 - Cx\ 4 * Cx\ a * Cx\ c1}}{Cx\ 2 * Cx\ a}$ $\vee$

$x = \dfrac{-Cx\ b1 - \sqrt{Cx\ b1\ pow\ 2 - Cx\ 4 * Cx\ a * Cx\ c1}}{Cx\ 2 * Cx\ a}$)

A8 : $0 < b1\ pow\ 2 - 4 * a1 * c1 \wedge$

($a1 < 0 \wedge$ ($b1 < \sqrt{b1\ pow\ 2 - 4 * a1 * c1}$ $\vee$

$\sqrt{b1\ pow\ 2 - 4 * a1 * c1} < -b1$) $\vee$

($0 < a1 \wedge$ ($\sqrt{b1\ pow\ 2 - 4 * a1 * c1} < b1$ $\vee$

$-b1 < \sqrt{b1\ pow\ 2 - 4 * a1 * c1}$)

$\Rightarrow$ first_order_cond ($\lambda x.\ Cx\ a * x\ pow\ 4 + Cx\ b * x\ pow\ 3 + Cx\ c * x\ pow\ 2 + Cx\ d$ $* x + Cx\ e$) x

In Theorem 4.11, Assumptions A3–A6 formally specify the relationship between the factors of the third order polynomial which is obtained from the derivative of the profit function in the conclusion of the theorem. The formal verification involves taking derivative of the cost function which reduces the order of the polynomial to three, and then using Theorem 3.5 to formally specify the factors of the function. This allows using Assumption A7 to wrap up the formal proof of the theorem. Whereas, Assumption A8 results from the condition of real nature of the roots for the real type of the profit function.

The proposed logical framework for the microeconomics behavioral modeling allowed to formally specify and verify the cost and utility models and profit-maximization analysis based on the polynomial functions upto the fourth order. The formal models, i.e., Definitions 4.1–4.6, and formally verified results, i.e., Theorems 4.1–4.2, provide the foundational support for the formal verification of cost and utility modeling and profit maximization using polynomials upto the fourth order, i.e., Theorems 4.3–4.11. The presence of an exhaustive set of assumptions provide a priori conditions on the domain and coefficients of the polynomial functions, which, in turn, can be very useful in the design of the cost and utility functions in real-world problems.

The core formalization of cost and utility modeling, i.e., Definitions 4.1–4.6 and Theorems 4.1–4.6, is applicable to microeconomics models based on the different mathematical functions such as logarithmic and exponential. The mechanization of the behavioral modeling involved considerable effort in building proof steps for the foundational results of the proposed methodology. The HOL Light proof script contains 2000 lines of code, which is available for the download at [9], and is equivalent to 250 man hours of development time.

4.5 Electricity Market Applications

Microeconomics modeling is used to solve optimal power flow problems, DR problems of designing incentives and spot pricing in electricity wholesale markets. In this section, we use our proposed formalization to formally verify the cost modeling of a thermal power plant [10] and utility modeling for consumers in the smart grid environment to maximize the financial gains in consuming electric power [5] . The case studies have employed quadratic polynomial type of the functions for cost and utility models to solve the economic problems in a power grid.

4.5.1 Quartic Polynomial Cost Function for Thermal Power Plants

A metaheuristic algorithm, i.e., artificial ant bee colony (ABC) algorithm, is applied to estimate the parameters of cost curves of distributed power generation from thermal power plants to reduce the operational cost of energy generation [10]. The ABC algorithm is used to estimate the curves for coal, oil and gas fuels using a quadratic cost function. We formally verify the estimated parameters of the quadratic type of cost function using our proposed formalization. The quadratic cost function (FC) for three fuel curves was defined as [10]

$$FC_j(P_{G_j}) = a_{oj} + a_{1j} P_{G_j} + a_{2j} P_{G_j}^2 + r_j \qquad (4.1)$$

Equation 4.1 is quadratic in variable P_{G_j}, which represents the amount of the generated power. where a_{ij}'s are coefficients of the jth quadratic polynomial and r is the error with jth generator. Thermal power plants have distributed generation capacity of five units with 10, 20, 30, 40 and 50 MW. An actual data, presented in Table 4.1, of the fuels cost was used to estimate the coefficients of the cost functions with the help of ABC algorithm [11].

The estimated parameters for quadratic type of cost curve are formally verified in HOL Light theorem prover as

Table 4.1 Estimated coefficients of cubic cost function using the ABC algorithm.

Coefficients	a_0	a_1	a_2
Estimated values (coal)	0.0414	7.5874	96.6046
Estimated values (oil)	0.0442	7.8779	101.5360
Estimated values (gas)	0.0439	8.0991	101.8179

Theorem 4.12 $\vdash \forall$ x.
A1 : (x $\in$ (0, 150))
$\Rightarrow$ 0 < real_derivative (Cx 0.0414 * x pow 2 + Cx 7.5874 * x + Cx 96.6046) $\wedge$
0 < real_derivative (real_derivative (λx. Cx 0.0414 * x pow 2 + Cx 7.5874
* x +
Cx 96.6046) $\wedge$
0 < real_derivative (Cx 0.0442 * x pow 2 + Cx 7.8779 * x + Cx 101.5360) $\wedge$
0 < real_derivative (real_derivative (λx. Cx 0.0442 * x pow 2 + Cx 7.8779
* x +
Cx 101.5360) $\wedge$
0 < real_derivative (Cx 0.0439 * x pow 2 + Cx 8.0991 * x + Cx 101.8179) $\wedge$
0 < real_derivative (Cx 0.0439 * x pow 2 + Cx 8.0991 * x +
Cx 101.8179)

The above theorem is formally verified using Theorem 4.3. Due to the universally quantified nature of Theorem 4.3., the proof in the HOL Light was conducted using just a few lines of the code. Theorem 4.12 explicitly states all the conditions on the derivatives of the quadratic cost function to ensure that the cost function satisfies all the modeling conditions encapsulated in Definition 4.3.

4.5.2   Quadratic Utility Function for Smart Grids

Microeconomics concept of utility modeling has been used to model the consumers' utility to design real-time pricing demand response strategy in a smart grid environment [5]. A quadratic polynomial type of function is employed to model the utility of consumers $(U(w, x))$, i.e.

$$U(\omega, x) = \begin{cases} \omega x - \frac{\alpha}{2}x^2 & 0 \leq x \leq \frac{\omega}{\alpha} \\ \frac{\omega}{\alpha} & \frac{\omega}{\alpha} < x \end{cases} \tag{4.2}$$

The above utility model is a function of two variables, i.e., ω and x, where ω is used to differentiate the user types in a smart grid such as household or an industrial, on the other hand, x represents electricity consumption of the particular user and α is a predetermined constant.

Utility function (4.2) has also been used to design a welfare objective function in a smart grid [5], i.e.

$$W(\omega, x) = U(\omega, x) - Px \tag{4.3}$$

$W(w, x)$ is a function of w and x, whereas P is the price of the electricity and x is the consumption measure in MW. First-order condition is used to find the maximum of the welfare objective function, (4.3), for a consumer, i.e.

$$\frac{d}{dx}W(\omega, x) = 0 \tag{4.4}$$

We use $w = 4$ and $\alpha = 0.5$, in the range of $(0, \frac{\omega}{x})$, for a single consumer to formally verify the utility modeling in HOL Light as

Theorem 4.13 $\vdash \forall$ x.
A1 : (x ∈ (0 , 8))
$\Rightarrow$ 0 ≤ real_derivative (λx. −Cx 0.25 ∗ x pow 2 + Cx 8 ∗ x) ∧
real_derivative (real_derivative (λx. −Cx 0.25 ∗ x pow 2 +
Cx 8 ∗ x ≤ 0)

The above theorem is formally verified mainly using Theorem 4.6. It is required to formally verify the conditions on the given quadratic cost function in Theorem 4.6. Additionally, assumption A1 explicitly specifies the energy consumption bound on the individual consumer in smart grid network.

Next, we have formally verified the maximization of utility objective function which is based on the quadratic utility function, i.e., (4.2), within the sound core of HOL Light as

Theorem 4.14 $\vdash \forall$ P x.
A1 : (x = 2∗P − 8)
$\Rightarrow$ (real_derivative (P∗x − (−Cx 0.25 ∗ x pow 2 + Cx 8 ∗ x)) = 0)

This theorem is formally verified mainly using Theorem 4.9. Whereas, Assumption A1 is the necessary condition for welfare maximization in terms of electricity price, p, and for the corresponding energy consumption, x. The formal verification code for Theorems 4.13 and 4.14 is composed of a few lines due to the formally verified generic results presented in Sects. 4.4.2 and 4.4.3.

4.6 Summary

In this chapter, we presented a theorem proving based methodology to formally analyze and verify the microeconomics behavioral models. We formalized strict convex or concave theory to formally model the cost and utility functions in an economy. To formally reason about the microeconomics models, we also formally verify the mathematical properties of the strict convex or concave functions and derivative tests for verifying the strict convexity or concavity properties. We further extended the logical framework by incorporating the profit, cost and utility models using polynomials upto the fourth order. To illustrate the usefulness of the proposed formalization, we formally verified the profit maximization, cost and utility models used in the economic dispatch and pricing algorithms in smart grids. This resulted in the explicit availability of an exhaustive set of the assumptions in the formally verified theorems and can be very useful in designing microeconomics models aimed at maximizing the potential financial benefits in a smart grid network.

References

1. J.M. Henderson, R.E. Quandt, et al., *Microeconomic Theory: A mathematical Approach* (1971)
2. A. Mas-Colell, M.D. Whinston, J.R. Green et al., *Microeconomic Theory*, vol. 1 (Oxford University Press, New York, 1995)
3. D.L. Debertin, *Applied Microeconomics: Consumption* (Production and Markets. CreateSpace Independent Publishing Platform, Scotts Valley, California, United States, 2012)
4. H. Bessembinder, M.L. Lemmon, Equilibrium pricing and optimal hedging in electricity forward markets. J. Finance **57**(3), 134–1382 (2002)
5. P. Samadi, A.H. Mohsenian-Rad, R. Schober, V.W. Wong, J. Jatskevich, Optimal realtime pricing algorithm based on utility maximization for smart grid, in *2010 First IEEE International Conference on Smart Grid Communications* (IEEE, 2010), pp. 415–420
6. Q. Wang, X. Liu, J. Du, F. Kong, Smart charging for electric vehicles: A survey from the algorithmic perspective. IEEE Commun. Surv. Tutorials **18**(2), 1500–1517 (2016)
7. F.A. Wolak, Identification and estimation of cost functions using observed bid data: an application to electricity markets (2001)
8. M. Fahrioglu, F.L. Alvarado, Using utility information to calibrate customer demand management behavior models. IEEE Trans. Power Syst. **16**(2), 317–322 (2001)
9. A. Ahmed, Formal beahvioral modeling in Microeconmics models. http://save.seecs.nust.edu.pk/projects/fcumm/. Accessed Mar 2021
10. Y. Sönmez, Estimation of fuel cost curve parameters for thermal power plants using the ABC algorithm. Turk. J. Electr. Eng. Comput. Sci., **21**(Sup. 1), 1827–1841 (2013)
11. M. El-Hawary, S. Mansour, Performance evaluation of parameter estimation algorithms for economic operation of power systems. IEEE Trans. Power Appar. Syst. **574–582** (1982)
12. X. Liu, P. Lu, B. Pan, Survey of convex optimization for aerospace applications. Astrodynamics **1**(1), 23–40 (2017)
13. Z. Qiu, G. Deconinck, R. Belmans, A literature survey of optimal power flow problems in the electricity market context, in *2009 IEEE/PES Power Systems Conference and Exposition* (IEEE, 2009), pp. 1–6
14. C. Niculescu, L.E. Persson, *Convex Functions and Their Applications* (Springer, Berlin, 2006)
15. Maxima. http://maxima.sourceforge.net/l. Accessed Mar 2021

Chapter 5
Formalization of Asymptotic Notations

Many of the smart grids objectives, such as cost reduction and power mitigation, depend upon the integration of plug-in electric vehicles and successful implementations of DR programs. In this context, asymptotic notations are quite frequently used to analyze and design low-computational algorithms. Traditionally, the analysis is conducted using the paper-and-pencil proof methods using notions of limits to model the asymptotic behaviors. However, paper-and-pencil methods are error prone due to human involvement. Whereas, the actual computational complexity of online algorithms, in smart grids, is crucial to achieve the objectives of cost reduction, power mitigation, reliability and quality of services. Considering the mission-critical application of smart grids, in this chapter, we develop a theorem proving based formalization to conduct formal asymptotic analysis of algorithms. We use the proposed formalization to formally verify Online cooRdinated CHARging Decision (ORCHARD) and online Expected Load Flattening (ELF) algorithm for plug-in electric vehicles.

5.1 Introduction

In this chapter, we present a framework for the formal asymptotic analysis of algorithms. We develop a real theory-based formalization of all asymptotic notations using the HOL Light theorem prover and utilize them to analyze the online scheduling algorithms for PEVs' charging. The proposed formalization includes the description of the asymptotic notations and formal verification of their properties in the higher-order logic theorem prover HOL Light. Moreover, this formalization is utilized to formally analyze the computational complexity of two online scheduling algorithms for PEVs' charging, i.e., the Online cooRdinated CHARging Decision (ORCHARD)

A. Ahmed et al., *Formal Analysis of Future Energy Systems Using Interactive Theorem Proving*, SpringerBriefs in Applied Sciences and Technology, https://doi.org/10.1007/978-3-030-78409-6_5

algorithm [1] and the Expected Load Flattening (ELF) algorithm [2]. Besides the formal verification of asymptotic properties in the sound core of HOL Light, the main challenge involved in the proposed work is the formulation of the mathematical problem related to the complexity of the given algorithms to conduct their formal analysis in the HOL Light theorem prover.

The rest of this chapter is organized as follows: Section 5.3 describes the formalization of asymptotic notations in HOL Light. Section 5.4 describes the formal analysis of the chosen algorithms in HOL Light. Finally, Sect. 5.5 concludes the chapter.

5.2 Proposed Methodology

In this section, we present the main steps of our proposed higher-order logic-based methodology to conduct the formal asymptotic analysis of online scheduling algorithms for PEVs' charging/discharging. Figure 5.1 presents the main steps of our proposed methodology for the formal asymptotic analysis and formal verification

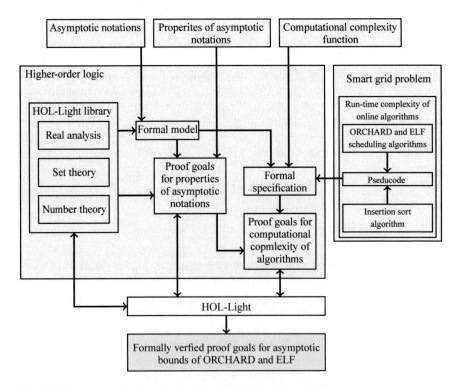

Fig. 5.1 Theorem proving based methodology for asymptotic analysis

of the computational complexity of online scheduling algorithms for PEVs' charging/discharging in smart grids.

1. In the first step, we develop a higher-order logic representation of asymptotic notions (Table 1.2) and formally verify the properties of the asymptotic notations. This allows us to formally reason and verify the run-time complexity of algorithms in the sound core of the HOL Light theorem prover.
2. The second step involves the representation of the ORCHARD and ELF algorithms as pseudocodes. As mentioned in Sect. 1.5, the pseudocodes provide computational complexity functions which are then used to assess the asymptotic behaviors of the algorithms. It is noteworthy that, such preprocessing steps are usually not available and contain important information. For example, to conduct the formal asymptotic behavior of PEV scheduling algorithms, we first formally verify the computational complexity of an ancillary algorithm, i.e., an insertion sort algorithm.
3. Finally, using the above formalization, we formally verify the asymptotic bounds for ORCHARD [1] and ELF [2] scheduling algorithms for PEVs in smart grids within the sound core of HOL Light theorem prover.

5.3 Formalization of Asymptotic Notations in HOL Light

The Big-O asymptotic notation, described in Table 1.2, is formally modeled in higher-order logic as

Definition 5.1 $\vdash \forall$ g. BigO (g : num $\rightarrow$ real) =

{(f : num $\rightarrow$ real)|($\exists$ c n_0. ($\forall$ n. n_0 $\leq$ n
$\Rightarrow$ 0 < c $\wedge$ 0 $\leq$ f(n) $\wedge$ f(n) $\leq$ c $*$ g(n)))}

In Definition 5.1, BigO is a higher-order logic function that accepts a function, g : num $\rightarrow$ real, and returns a set of functions, f : num $\rightarrow$ real, which are bounded by the growth rate of c $*$ g(n), i.e., 0 $\leq$ f(n) $\leq$ c $*$ g(n). Whereas c is a strictly positive real constant, i.e., 0 < c. The size of the input is modeled using a variable of natural type, n : (num), which is bounded by a constant natural number, n_0.

Definition 5.2 $\vdash \forall$ g. BigOmega (g : num $\rightarrow$ real) =

{(f : num $\rightarrow$ real)|($\exists$ c n_0. ($\forall$ n. n_0 $\leq$ n
$\Rightarrow$ 0 < c $\wedge$ 0 $\leq$ c $*$ g(n) $\leq$ f(n)))}

Definition 5.2 formally models the BigOmega asymptotic notation which differs from Definition 5.1 only in the inequality relating the rate of growths of the given function, g : num $\rightarrow$ real, and every member function, f : num $\rightarrow$ real, of the set, i.e., 0 $\leq$ c $*$ g(n) $\leq$ f(n).

Definition 5.3 $\vdash \forall$ g. BigTheta (g : num $\rightarrow$ real) =
$\{(\,f : num \rightarrow real\,)\,|\,(\exists$ c1 c2 n_0. $(\forall$ n. n_0 $\leq$ n
$\Rightarrow$ 0 < c1 $\wedge$ 0 < c2 $\wedge$ 0 $\leq$ c1 $*$ g(n) $\leq$ f(n) $\leq$ c2 $*$ g(n)))$\}$

Definition 5.3 models the BigTheta asymptotic notation which uses two real constants, c1 and c2, to relate the growth rates of the given function, g : num $\rightarrow$ real, and every member function, f : num $\rightarrow$ real, of the set, i.e., c1 $*$ g(n) $\leq$ f(n) $\leq$ c2 $*$ g(n). Moreover, the constants are strictly positive, i.e., 0 < c1 and 0 < c2.

Definition 5.4 $\vdash \forall$ g. LittleO (g : num $\rightarrow$ real) =
$\{(\,f : num \rightarrow real\,)\,|\,(\exists$ c n_0. $(\forall$ n. n_0 $\leq$ n
$\Rightarrow$ 0 $\leq$ c $\wedge$ 0 < f(n) $\wedge$ f(n) < c $*$ g(n)))$\}$

Definition 5.4 formally models the LittleO asymptotic notation using a real constant c to form a set that contains all the functions with growth rates strictly less than the growth rate of c $*$ g(n). Due to strict inequality, LittleO is used to specify strict asymptotic bound complexity of algorithms.

Definition 5.5 $\vdash \forall$ g. LittleOmega (g : num $\rightarrow$ real) =
$\{(\,f : num \rightarrow real\,)\,|\,(\exists$ c n_0. $(\forall$ n. n_0 $\leq$ n
$\Rightarrow$ 0 < c $\wedge$ 0 $\leq$ c $*$ g(n) < f(n)))$\}$

Definition 5.5, formally models the LittleOmega asymptotic notation which differs from Definition 5.4 only in the inequality relating the rate of growths of the given function, g : num $\rightarrow$ real, and every member function, f, of the set, i.e., c $*$ g(n) < f(n).

The formal modeling of asymptotic notations allow to reason and verify the properties of the asymptotic notations in the sound core of the HOL Light theorem prover.

5.3.1 Formal Verification of Asymptotic Notations' Properties

In this section, we formally verify the properties of the asymptotic notations which will be later useful in formally reasoning the asymptotic bounds of the algorithms.

The transitivity property of the asymptotic notations is formally verified in HOL Light as

Theorem 5.1 $\vdash \forall$ f g h., f $\in$ BigO g $\wedge$ g $\in$ BigO h $\Rightarrow$ f $\in$ BigO h

Theorem 5.2 $\vdash \forall$ f g h.
            f $\in$ BigOmega g $\wedge$ g $\in$ BigOmega h $\Rightarrow$ f $\in$ BigOmega h

Theorem 5.3 $\vdash \forall$ f g h.
            f $\in$ BigTheta g $\wedge$ g $\in$ BigTheta h $\Rightarrow$ f $\in$ BigTheta h

Theorem 5.4 $\vdash \forall$ f g h.
   f $\in$ Littelo g $\land$ g $\in$ Littelo h $\Rightarrow$ f $\in$ Littelo h

Theorem 5.5 $\vdash \forall$ f g h.
   f $\in$ Littelomega g $\land$ g $\in$ Littelomega h $\Rightarrow$ f $\in$ Littelomega h

Theorems 5.1–5.5 formally verify the transitivity properties of Big-O, Big-Ω, Big-Θ, Little-o and Little-ω notations. The formal verification is conducted using formal definitions of asymptotic notations, i.e., Definitions 5.1–5.5, and using the existential quantifier to provide an evidence to complete the formal proofs.

Next, we formally verify the reflexivity property of the asymptotic notations in HOL Light theorem prover as

Theorem 5.6 $\vdash \forall$ f. $(\exists n_0. \ (\forall$ m. $n_0 \leq$ m $\Rightarrow 0 \leq$ f m$) \Rightarrow$ (f $\in$ BigO f))

Theorem 5.7 $\vdash \forall$ f. $(\exists n_0. \ (\forall$ m. $n_0 \leq$ m $\Rightarrow 0 \leq$ f m$) \Rightarrow$ (f $\in$ BigOmega f))

Theorem 5.8 $\vdash \forall$ f. $(\exists n_0. \ (\forall$ m. $n_0 \leq$ m $\Rightarrow 0 \leq$ f m$) \Rightarrow$ (f $\in$ BigTheta f))

Theorems 5.6–5.8 formally verify that any function f is asymptotically related to itself for Big-O, Big-ω and Big-Θ notations. The formal verification of the reflexivity property exploits the equality involved in the definitions of these asymptotic notations.

Now, we formally verify the summation property of the Big-O and Big-Θ in the HOL Light theorem prover as

Theorem 5.9 $\vdash \forall$ t1 t2 g1 g2.
   (t1 $\in$ BigO g1) $\land$ (t2 $\in$ BigO g2)
                   $\Rightarrow$ (λn. t1 n + t2 n) $\in$ (BigO (max(g1 , g2))

Theorem 5.10 $\vdash \forall$ t1 t2 g1 g2.
   (t1 $\in$ BigOmega g1) $\land$ (t2 $\in$ BigOmega g2)
                   $\Rightarrow$ (λn. t1 n + t2 n) $\in$ (BigOmega (min(g1 , g2))

Theorem 5.9 formally verifies that for two functions t1 and t2, which have BigO asymptotic relationship with g1 and g2, respectively, the summation of these two functions will also be in the BigO of maximum of the g1 and g2 functions. Similarly, Theorem 5.10 formally verifies that for two functions t1 and t2 having the BigOmega asymptotic relationship with g1 and g2, respectively, the summation of these two functions will also be in the BigOmega of minimum of the g1 and g2. It is due to the fact that BigO is upper asymptotic bound and BigOmega is lower asymptotic bound.

The symmetry property for Big-Ω and Big-Θ are formally verified as

Theorem 5.11 $\vdash \forall$ f g.   f $\in$ BigOmega g $\Rightarrow$ g $\in$ BigOmega f

Theorem 5.12 $\vdash \forall$ f g.   f $\in$ BigTheta g $\Rightarrow$ g $\in$ BigTheta f

Theorem 5.11 formally verifies that, if the growth rate of the function f is related to a function g through BigOmega, then the growth rate of g will also be related to f through BigOmega. Similarly, Theorem 5.12 formally verifies the symmetric relationship for BigTheta asymptotic notation.

The transpose symmetry property of Big-O and Little-o is formally verified in the HOL Light theorem prover as

Theorem 5.13 $\vdash \forall$ f g.   f $\in$ BigO g $\Rightarrow$ g $\in$ BigOmega f

Theorem 5.14 $\vdash \forall$ f g.   f $\in$ Littelo g $\Rightarrow$ g $\in$ LittleOmega f

Theorem 5.13 formally verifies that, if f has a BigO relationship with function g, then g will also be in the BigOmega of f. Similar relationship is formally verified for Littelo and Littel Omega in Theorem 5.14.

The Big-O notation plays a crucial role in the formal verification of the online scheduling PEV algorithms, therefore, we present two related properties of the Big-O notation in the HOL Light theorem prover as

Theorem 5.15 $\vdash \forall$ f g.   f $\in$ BigO g $\Rightarrow$ $\forall$ k. (λn. k $*$ f n) $\in$ (BigO g)

Theorem 5.15 formally verifies that the BigO relationship between two functions f and g is not affected by the constant multiplication, k.

Theorem 5.16 $\vdash \forall$ t1 t2 g1 g2.
  (t1 $\in$ BigO g1) $\wedge$ (t2 $\in$ BigO g2)
      $\Rightarrow$ (λn. t1 n $*$ t2 n) $\in$ (BigO (g1 $*$ g2))

Theorem 5.16 formally verifies the multiplication of two functions, i.e., t1 and t2, will be in BigO of g1$*$g2, given that t1 and t2 are related to g1 and g2 through BigO notation, independently.

Theorems 5.1–5.16 provide a higher-order logic framework to conduct formal asymptotic analysis of algorithms. In the next section, we use this logical framework to formally analyze the asymptotic bounds of the online PEV scheduling algorithms.

5.4 Formal Asymptotic Analysis of Scheduling Algorithms for PEVs

In this section, we use our proposed methodology to formally verify the worst-case computational complexity of two case studies, i.e., Online cooRdinated CHARging Decision (ORCHARD) [1] and online Expected Load Flattening (ELF) algorithm [2].

5.4.1 Formal Analysis of Insertion Sort Algorithm

Sorting is a logical procedure to arrange a given list of items in specific or desired order, such as descending, ascending, alphabetical, chronological or topological, depending on the given data type. Sorting has ubiquitous applications in the computer programs as a preprocessing or even main logical step, such as information retrieval, crunching, searching and data mining. Therefore, many algorithm designs have been proposed, e.g., insertion sort, heapsort, merge sort, counting sort, quicksort and radix sort, to efficiently sort the data [3]. Insertion sort algorithm has polynomial time complexity and its worst-case running time is bounded by a quadratic polynomial, i.e., $O(n^2)$, with respect to the size of the input n [3]. For a given array of the unsorted data, the insertion sort algorithm incrementally sorts the data of the array over the length of the array, N. In each iteration, it selects k ($k > 1$) entries of the given array and sorts them in a desired order. The variable k is incremented until $k = N$ to completely sort the given array.

Algorithm 1 Insertion Sort	Cost	Time Steps
Input: list A		
1: **for** $j = 2$ **to** n	c_1	n
2: $key = A[j]$	c_2	$n - 1$
3: $i = j - 1$	c_3	$n - 1$
4: **while** $i > 0$ **and** $A[i] > key$	c_4	$\sum_{j=2}^{n} t_j$
5: $A[i + 1] = A[i]$	c_5	$\sum_{j=2}^{n}(t_j - 1)$
6: $i = i - 1$	c_6	$\sum_{j=2}^{n}(t_j - 1)$
7: $A[i + 1] = key$	c_7	$n - 1$

Algorithm 1 represents the pseudocode of an insertion sort algorithm for sorting an input data array, A. The algorithm uses `for` loop to conduct the sorting of the array A in Line 1. The temporary variable, j, of the loop is initiated with the value of 2 as sorting requires at least two elements for the comparison purpose. This loop considers all the elements of the given array one-by-one, and therefore, requires n time steps, with an associated cost of c_1 for each step. In the next line, current entry of the array A is saved to a temporary variable key which will enable comparison operation among the elements of the subarry $A[1...j]$. Line 3 uses another temporary variable i to save the value of $j - 1$, which is, in fact, the number of the comparisons the insert sort algorithm has to conduct for sorting the subarray $A[1...j]$. Again, the number of the comparisons are one less than the length of the subarray A. In Line 4, the algorithm uses a `while` loop to compare the elements of the subarray A. `While` loop has two conditions in conjunction, where $i > 0$ ensures that the subarray is not an empty list and $A[i] > key$ is used to compare the ith entry of the subarray with the key entry of the array. The algorithm compares all of the elements of subarray with *key* sequentially, and therefore, $\sum_{j=2}^{n} t_j$ operations are required, with an associated cost of $c4$ for each operation, to sort a subarray A. The next two lines, i.e., Lines 5 and 6, make the body of the `while` loop. In Line 5, a swap of the entries of subarray $A[1..i]$ is perforemed, if the conditions of the `while` loop are *true*. The conditions, $A[i] > key$, and $A[i + 1] = A[i]$, are used to sort the array A in the ascending order. Next, the program statement in Line 6 decrements the index of the subarray A for the next iteration of the `While` loop. Whereas c_5 and c_6 are the associated costs for each operation in Lines 5 and 6, respectively. Finally, Line 7 updates the location of the *key* in the subarray A. This operation is executed $n - 1$ times due to the initialization of $j = 2$. The computational complexity function, $T(n)$, for an insertion sort algorithm from the pseudocode (1) is obtained by multiplying the cost, c_i, and number of the time steps for each statement, i.e.

$$T(n) = c_1 n + c_2(n - 1) + c_3(n - 1) + c_4 \sum_{j=2}^{n} t_j$$

$$+ c_5 \sum_{j=2}^{n}(t_j - 1) + c_6 \sum_{j=2}^{n}(t_j - 1) + c_7(n - 1)$$

For the worst-case scenario, i.e., considering maximum number of operations for the execution of the task, we can rewrite $T(n)$ as

$$T(n) = \left(\frac{c4}{2} + \frac{c5}{2} + \frac{c6}{2}\right)n^2 - \left(c_1 + c_2 + c_3 + \frac{c4}{2} + \frac{c5}{2}\right.$$

$$\left. + \frac{c6}{2} + c_7\right)n - (c_2 + c_3 + c_4 + c_7)$$

We formally model the complexity function, $T(n)$, of insertion sort algorithm using higher-order logic as

Definition 5.6 $\vdash \forall$ n c1 c2 c3 c4 c5 c6 c7.
  i_sort_wc_t n c1 c2 c3 c4 c5 c6 c7 =
    ($\frac{c4}{2}$ + +$\frac{c5}{2}$ + $\frac{c6}{2}$) n pow 2 − (c$_1$ + c$_2$ + c$_3$
      + $\frac{c4}{2}$ + $\frac{c5}{2}$ + $\frac{c6}{2}$ + c7)n − (c2 +c3 +c4 +c7)

In above definition, c_1 to c_7 are the real constants that represent the cost of the primitive operations in an insertion sort algorithm, as described in Algorithm 1, and n is an integer representing the length of the array A.

We use the formalization of Sect. 5.3 to formally verify the asymptotic behavior of the insertion sort algorithm as

Theorem 5.17 $\vdash \forall$ n c1 c2 c3 c4 c5 c6 c7.
   0 < n ∧   0 < c1 ∧ 0 < c2 ∧ 0 < c3 ∧
   0 < c4 ∧ 0 < c5 ∧ 0 < c6 ∧ 0 < c7
      ⇒ (λ n. i_sort_wc_t n c1 c2 c3 c4 c5 c6 c7 ∈ BigO n pow 2)

Theorem 5.17 formally verifies that the computational complexity function of insertion sort algorithm is polynomially bounded, i.e., $O(n^2)$, under the explicit conditions on the size of the array A, i.e., $0 < n$, and costs of the basic operations, i.e., $0 < c_i$.

The formalization of the insertion sort algorithm plays a vital role in formally verifying the computational complexity of the online scheduling algorithms for PEVs, as presented in the next sections.

5.4.2 Online cooRdinated CHARging Decision (ORCHARD)

Online cooRdinated CHARging Decision (ORCHARD) is a low-computational complexity algorithm to schedule the charging/discharging of PEVs in a smart grid [1]. The algorithm uses convex optimization theory [6] to mathematically formulate the PEV scheduling problem. Convex optimization has a distinguishing advantage of being a fairly complete theory, and hence provides an efficient and reliable optimal solution to the complicated task of scheduling PEVs in a smart grid. Moreover, to counter the randomness associated with the arrival and departure of PEVs, ORCHARD utilizes the speed scaling technique [4], which is commonly used in the computer and communication systems as a power management technique to reduce the energy consumption of the systems. This problem has remarkable resemblance with the scheduling of the PEVs, the arrival and departure time and allocation of resources to the processing task resembles the arrival, departure and power charging or discharging rates of the PEVs. Speed scaling problem was first presented by Yao et al [4] to schedule the processor tasks and they presented two online algorithms, namely, average rate (AR) and optimal available (OA). AR runs the given task of processing at an average speed, whereas OA schedules the newly arrived task assuming no task arrival in the future. The ORCHARD uses a variant of the OA algorithm, termed as qOA, which improves the efficiency of the OA algorithm. Moreover, ORCHARD employs an interior point method technique from the convex theory to solve the scheduling problem of the PEVs. However, the interior point method has an exponential computational complexity with respect to the size of the input, and therefore, ORCHARD uses a Low-computational complexity routine to solve the convex optimization problem of PEVs.

The considered low-computational complexity routine finds its inspiration from the working principle of the interior point method. Like the interior point method, Low-computational complexity routine tries to balance the workload by balancing the total workload for an optimal solution. The considered low-computational complexity routine defines workload in terms of the interval density and uses this measure to shift load among the adjacent intervals to balance the workload in order to achieve an optimal solution for PEV scheduling. This results in an optimal solution to the scheduling problem of the PEVs in an online manner.

As worst-case scenario is the most critical scenario, so we consider the worst-case scenario of ORCHARD, i.e., scheduler has N number of PEVs at a given time t. The Low-computational complexity algorithm needs to schedule N PEVs with

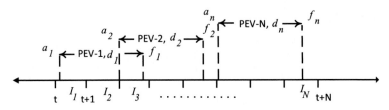

Fig. 5.2 An online PEV charging scheduling instance at any time t

a portfolio containing the information of their arrival (a_i), departure time (f_i) and charging demand (d_i), in terms of charging rate. The Low-computational complexity routine balances the workload, w_i, among N intervals, corresponding to N PEVs. In the worst-case scenario, Low-computational complexity routine will have to search for N windows at every time step of the time-horizon of length N, which amounts to N^2 windows of the time, as shown in Fig 5.2.

Algorithm 2 describes the pseudocode for the Low-computational routine. Input to the scheduler is contained in the two sets, P and W. Set P represents the candidate PEVs for scheduling, whereas set W contains information regarding the N^2 time windows. To schedule a certain time window, τ, set W contains density of workload, $\rho_i = \frac{1}{\delta_\tau} \sum_{P_i \in S_\tau} d_i$, of the time intervals comprising the time window with length δ_τ. Whereas S_τ is a subset of P, which contains the PEVs for the current time slot, τ. For N PEVs at a given time, the while loop executes for N times. Lines 2 and 3 together describe the Low-computational complexity routine for scheduling. In Line 2, a certain time interval τ with highest work is selected based on the workload density, ρ. This time interval is sorted using insertion sorted and, in Line 3, it allocates a charging rate to all PEVs, such that the density of the workload is optimal. As we have formally modeled and verified the computational complexity of the insertion sort algorithm, therefore, the two steps will amount to computational complexity of $O((N^2)^2)$, where $n = N^2$ in this case. In Lines 4 and 5, the Low-complexity routine updates the PEV set and time-horizon for the next iteration of while loop. However, the two operations do not depend upon the size of the input, and therefore, we do not incorporate them in our computational complexity function for Low-computational routine. We model the run-time function in HOL Light as

Definition 5.7 ⊢ ∀ N c1 c2 c3 c4 c5 c6 c7.
    lcr_wc_t N c1 c2 c3 c4 c5 c6 c7 =
        N * i_sort_wc_t N pow 2 c1 c2 ,c3 c4 c5 c6 c7

In the above definition, `lcr_wc_t` is a higher-order logic function which accepts the operation costs, i.e., c_1 to c_7, and number of PEVs, i.e., N. `lcr_wc_t` uses the formal definition of the worst-case computational complexity of the insertion sort to model the `BigO` asymptotic bound of the ORCHARD. Notably, we use N^2 in the higher-order logic function `i_sort_wc_t` to model the computational complexity of the Low-complexity routine in Algorithm 2.

Algorithm 2 Low-complexity Routine   Worst-case

Input: $P := P_1, ..., P_N, \tau, \rho_\tau = \frac{1}{\delta_\tau} \sum_{P_i \in S_\tau} d_i, S_\tau \in P$

1: **while** $P \neq \{\}$ N

2:   Determine the time interval , τ of the
     highest intensity, i.e., ρ_τ $O(N^4)$

3:   Allocate the charging rate, $\widehat{x_i}$, to all PEVs
     such that $\rho_\tau = \sum_{P_i \in S_\tau} \widehat{x_i}$

4:   Set $P := P \backslash S_I$

5:   Remove I from the time horizon
     and update the departure, arrival
     and residual demands

We formally verify the worst-case complexity for the ORCHARD algorithm in HOL Light using Definitions 5.1, 5.6, 5.7, Theorem 5.17 and formally verified properties of the Big-O notation, presented in Sect. 5.3 as

Theorem 5.18 $\vdash$ $\forall$ N c1 c2 c3 c4 c5 c6 c7.
  $0 < $ N $\wedge$ $0 < $ c1 $\wedge$ $0 < $ c2 $\wedge$ $0 < $ c3
      $\wedge$ $0 < $ c4 $\wedge 0 < $ c5 $\wedge 0 < $ c6 $\wedge$ $0 < $ c7
            $\Rightarrow$ lcr_wc_t N c1 c2 c3 c4 c5 c6 c7 $\in$ BigO N pow 5

Formal verification using theorem proving based methodology resulted in the explicit set of assumptions on all the variables of the asymptotic bound of the ORCHARD algorithm. These assumptions on the variables corresponds to the important physical constraints such as $0 < N$ to ensure the correct execution of the algorithms.

5.4.3   Low-complexity online Expected Load Flattening (ELF) algorithm

A model predictive control theory-based approach has been used to design a low complexity algorithm, i.e., Expected Load Flattening (ELF), for scheduling the PEVs in a smart grid [2]. Model predictive control theory is the state-of-the-art method, which mainly relies upon the process model and statistics of the data to optimally forecast the future state of the process using a feedback mechanism. This feature allows MPC to cater the potential uncertainties such as load variances arising due to the large scale integration of PEVs in smart grids.

ELF working principle also exploits the structure of optimal solution. Thus, MPC forecasts are aimed at flattening the demand curve of the PEVs at every time step. The ELF algorithm divides the time-horizon in T equal length time intervals and schedules the randomly arriving and departing PEVs using demands of PEVs (d_i), unfinished demands of PEVs ($\hat{d}_i^k$) and expected charging demands (ξ_t), at every time

step, illustrated in Fig 5.5. The unfinished charging demands of PEVs, $\hat{d}_i^k$, which
have pluged-in at time k is defined using the charging demands, d_i, and charging
rates, x_{it}, i.e., $\hat{d}_i^k = d_i - \sum_{t=a_i}^{k-1} x_{it}$. Whereas the demand of PEVs which have not
pluged-in until the time step $k-1$ is assumed to be the equal to the profile demand,
i.e., $d_i^k = d_i$. A state vector D_t is used to model the total electricity demand at any
time over the time-horizon T

$$\mathbf{D}_t = [l_t, \tilde{d}_t^t, \tilde{d}_{t+1}^t, ..., \tilde{d}_T^t] \tag{5.1}$$

The vector $\mathbf{D}_t$ represents the PEVs remaining demands, $\tilde{d}_i^t$, and electricity demand
except the PEVs, l_t. The ELF algorithm incorporates the future demands as a random
arrival events, i.e.

$$\xi_t = [l_t, \gamma_t^t, \gamma_{t+1}^t, ..., \gamma_T^t] \tag{5.2}$$

The above state vector represents the based load, l_t, and future demands needed to be
fulfilled, $\gamma_{t'}^t$, by time t' as random variables. The expected future demands of PEVs
are

$$\epsilon_t = [\alpha_t, \beta_t^t, \beta_{t+1}^t, ..., \beta_T^t] \tag{5.3}$$

Where, α and β are expected values of the random variables representing future
base load and demands beyond the current time step, i.e., $\alpha = E[l_i]$ and $\beta = E[\gamma_i]$,
respectively. The ELF algorithm uses (5.1) and (5.3) to define the state of the system

$$\bar{d}_{t''}^t = \begin{cases} \tilde{d}_{t''}^t, & \text{for } t'' = k, ..., T, t' = k \\ \beta_{t''}^t, & \text{fro } t'' = t', ..., T, t' = k+1, ..., T \end{cases} \tag{5.4}$$

The ELF algorithm utilizes (5.4) to schedule the PEVs by flattening the demand
curve of the energy demand by predicting optimal solutions, $s^*(t)$, at every time
step, as shown in Fig. 5.3.

Algorithm 3 describes the main logic of the ELF algorithm as a pseudocode. The
pseducode consists of the sequence of the operations which are applied by an ELF
algorithm to solve the PEV scheduling problem in a smart grid. The ELF algorithm

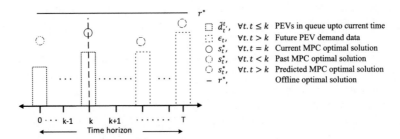

Fig. 5.3 MPC-Based Expected Load Flattening (ELF) algorithm

requires the total unfinished charging demands, $\mathbf{D_k}$, and expected charging demands, ϵ_t, at the current time, $t = k$, as an input. The charging schedule, as an optimal solution, s_k, is the output of the algorithm. In the first line, the global variables i and j are initialized with 0. A repeat $-$ until construct is used in Line 2 to implement the low complexity ELF routine. The two variables, i and j are initialized in Line 3 with the information of time-horizon. The variable i is used to index the time-horizon from the current time step onward until the total length of the time-horizon.

Algorithm 3 Expected Load Flattening (ELF)   Worst-case

Input: $\mathbf{D_k}, \epsilon_k, t = k + 1, ..., T$

Output: Charging rate, s_k, at time slot k

1: initialization $i = 0$ and $j = 0$

2: **repeat**

3:    For all time slots, $O(T^2)$
    $i = k, ..., T, j = i, ..., T$, calculate
    $$i^*, j^* = \arg \max_{k \le i \le j \le T} \{\frac{\sum_{t'=i}^{j}(\sum_{t''=t'}^{j} \bar{d}_{t''}^{t'} + \alpha_{t'})}{j - i + 1}\}$$

4:    set
    $$y^* = \{\frac{\sum_{t'=i}^{j}(\sum_{t''=t'}^{j} \bar{d}_{t''}^{t'} + \beta_{t'})}{j - i + 1}\}$$

5:    delete time slots i^*, j^* and relable the existing
    time slot $t > j^*$ as $t - j + i - 1$

6: **until** $i \ne k$ T

7: Set $s_k = y^* - l_k$

Whereas the variable j is defined in terms of the variable i to incorporate the information contained in the state vectors $\mathbf{D_k}$ and ϵ_k. In the next step, the algorithm sorts the maximum load density based on the total electricity load, including the remaining, expected and base load. The sorting is assumed to be conducted using the insertion sort algorithm, as was in the case of ORCHARD. The main reason for using insertion sort is that the claimed computational complexities of ORCHARD and ELF, in original contributions, were found to conform with the use of this choice. In Line 4, the result of the previous step, i.e., maximum charging demand, is saved in a local variable, y. In the next step, variables i and j are relabeled for the next iteration of the ELF algorithm. In Line 6, the condition for maximum charging demand index and current time step are compared to take the decision of scheduling the PEVs with an optimal charging rate. As we have considered worst-case scenario, therefore, the repeat $-$ until loop runs T times. The last operation, in Line 7, subtracts the base load from the solution obtained from the repeat-until loop for an optimal solution. As operations in Lines 4, 5 and 6 do not depend upon the size of input, therefore, the execution time of these operations is not considered in the complexity analysis of the ELF algorithm. We formally model the computational complexity of Algorithm 3 in HOL Light as

Definition 5.8 $\vdash$ $\forall$ T c1 c2 c3 c4 c5 c6 c7.
     elf_wc_t T c1 c2 c3 c4 c5 c6 c7 =
        T * i_sort_wc_t T pow 2 c1 c2 c3 c4 c5 c6 c7

Definition 5.8 is formally modeled using Definition 5.6, where T is the number of equal length intervals, defining the time-horizon of ELF, and c_i's are the constant multipliers representing the cost of the basic operations of the insertion sort algorithm. The function elf_wc_t is a multiplication of T and the computational complexity of insertion sort algorithm, where T^2 corresponds to the total number of comparisons in the ELF for T time intervals.

We formally verify the computational complexity of the ELF algorithm using the formalization of Big-O notation and formalization of insertion sort algorithm in HOL Light as follows:

Theorem 5.19 $\vdash$ $\forall$ T c1 c2 c3 c4 c5 c6 c7.
$0 <$ T $\wedge$ $0 <$ c1 $\wedge$ $0 <$ c2 $\wedge$ $0 <$ c3 $\wedge$ $0 <$ c4 $\wedge$ $0 <$ c5 $\wedge$ $0 <$ c6
      $\wedge$ $0 <$ c7 $\Rightarrow$ lcr_wc_t T c1 c2 c3c4 c5 c6 c7 $\in$ BigO (T pow 3)

Theorem 5.19 formally verifies the asymptotic bound of the ELF algorithm and provides all the necessary conditions on the variables such as T anc c_i's. These formally verified results with specifications can avoid errors in the implementation stages of the algorithms, and hence ensure reliable, secure and efficient integration of PEVs in a smart grid.

This chapter provides a foundational framework for the formal analysis and verification of the asymptotic bounds of algorithms which play a central role in the design of low computation algorithms for PEV scheduling in smart grids. The proposed methodology is employed to formally verify the asymptotic bounds of the ORCHARD and ELF PEV scheduling algorithms as Theorems 5.18 and 5.19 in the sound core of HOL Light theorem prover. The formal asymptotic analysis resulted in the explicit conditions on the complexity functions which can be valuable in the implementation of the corresponding algorithms in smart grids.

The foundational formalization of asymptotic notations can be used to formally analyze and verify the computational complexity of any algorithm. This makes the proposed formalization very useful for the formal verification of many safety or mission-critical applications of algorithm design. The proposed formalization of asymptotic notations can be viewed as a primary resource to formally verify the asymptotic bounds of algorithms based on the different algorithm design strategies, such as dynamic programming, Brute force, divide-and-conquer, etc. For example, the case studies of ORCHARD and ELF algorithms can be formally verified using merg sort or quick sort to formally verify and compare the performances of the algorithms. This kind of analysis is highly desirable in the development and implementation of the algorithms that are to be used in safety or mission-critical applications. The proposed formalization can be extended to incorporate more details such as explicit computational costs associated with the basic primitive operations in a PEV

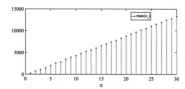

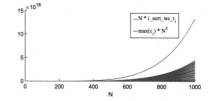

(a) Cost function of $n \leq 30$ schedulers.

(b) Asymptotic growth of $n \leq 30$ schedulers.

Fig. 5.4 Asymptotic behavior of the ORCHARD algorithm for increasing cost functions. **a** Cost function of $n \leq 30$ schedulers; **b** Asymptotic growth of $n \leq 30$ schedulers

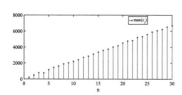

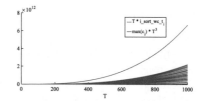

(a) Cost function of $n \leq 30$ schedulers.

(b) Asymptotic growth of $n \leq 30$ schedulers.

Fig. 5.5 Asymptotic behavior of the ELF algorithm for increasing cost functions. **a** Cost function of $n \leq 30$ schedulers; **b** Asymptotic growth of $n \leq 30$ schedulers

scheduling algorithm. Due to exhaustive nature of theorem proving, this approach may result in the necessary conditions in terms of the system parameters which are direly needed for the efficient implementation of the algorithms.

We illustrate the usefulness of the preconditions obtained from the proposed formalization for the analysis and verification of asymptotic behavior of the algorithms using MATLAB, which can be used at the early stages of the deployment or implementation of the online scheduling algorithms to avoid subtle errors which may lead to the insecure and unreliable smart grid operations. We consider 30 schedulers which have a fixed upper limit of the load, i.e., 1000 PEVs, with an increasing cost of the operations and model the computational complexity using the formalization of insertion sort, ORCHARD and ELF algorithms. We define the asymptotic bounds of the schedulers using ORCHARD and ELF bounds and model them using Definition 5.6. We randomly generated the cost of operations for the schedulers operating under the asymptotic bounds of ORCHARD and ELF. We used preconditions of Theorems 5.17 and 5.18 to model the asymptotic behavior of schedulers for the two algorithms. Figures 5.4 and 5.5 present the simulation results for the asymptotic behaviors of the schedulers utilizing ORCHARD and ELF computational complexity functions, respectively. Figures 5.4a and 5.5a represent the maximum cost functions for the two algorithms, whereas Figs. 5.4b and 5.5b describe the corresponding asymptotic behavior of the two algorithms for the worst-case scenario.

5.4.4   Simulation Results

The simulation results show the increasing cost of operations and corresponding upper bounds of the algorithms, i.e., Big-O asymptote. Figures 5.4a and 5.5a show increasing costs for 30 schedulers, whereas Figs. 5.4b and 5.5b show the corresponding behaviors of the computational complexities of ORCHARD and ELF algorithms. This type of the analysis can be really useful to design algorithms with a desired latency and quality of services in a PEV-integrated grid system.

5.5   Summary

In this chapter, we proposed a theorem proving based methodology to conduct asymptotic analysis of algorithms. The formalization consists of the formal modeling and verification of the asymptotic notations and their properties within the sound core of the HOL Light theorem prover. The formalization is used to formally verify two state-of-the-art low-computational complexity algorithms for PEV scheduling in smart grids, i.e., ORCHARD and ELF. This leads to the formal verification of the asymptotic bounds of insertion sort, ORCHARD and ELF algorithms, resulting in valuable preconditions. Furthermore, we demonstrated the use of the preconditions obtained from the formal analysis in the conventional algorithm design and analysis using a general purpose tool, i.e., MATLAB, by considering a PEV scheduler setup with ORCHARD and ELF computational complexities.

References

1. W. Tang, S. Bi, Y.J.A. Zhang, Online coordinated charging decision algorithm for electric vehicles without future information. IEEE Trans. Smart Grid **5**(6), 2810–2824 (2014)
2. W. Tang, Y.J.A. Zhang, A model predictive control approach for low-complexity electric vehicle charging scheduling: optimality and scalability. IEEE Trans. Power Syst. **32**(2), 1050–063 (2016)
3. T.H. Cormen, C.E. Leiserson, R.L. Rivest, C. Stein, *Introduction to Algorithms*, 2nd edn. (The MIT Press, 2001). http://web.ist.utl.pt/fabio.ferreira/material/asa/clrs.pdf
4. F. Yao, A. Demers, S. Shenker, A scheduling model for reduced cpu energy, in *Proceedings 36th Annual Symposium on Foundations of Computer Science* (IEEE, 1995), pp. 374–382
5. N. Bansal, H.L. Chan, K. Pruhs, D. Katz, Improved bounds for speed scaling in devices obeying the cube-root rule, in *Automata, Languages and Programming* (2009), pp. 144–55
6. S. Boyd, L. Vandenberghe, *Convex Optimization* (Cambridge University Press, Cambridge, UK, 2004)

Chapter 6
Conclusions

This book presented a higher-order logic theorem proving based approach for formally analyzing and verifying the safety and mission-critical aspects of smart grids. The main reason for the proposed approach is to overcome the limitations of traditional paper-and-pencil and simulation techniques, and ensure safe and secure grid operations. Higher-order logic theorem proving is a formal methods technique which allows to formally express and reason about the complex systems such as continuous systems.

6.1 Summary

This book provided a higher-order logic formalization of stability, microeconomics and algorithm design analysis. These generic formalizations are then used to formally verify some of the key design aspects at generation, transmission and distribution and energy and data processing sides in smart grids.

The work is presented in HOL Light theorem prover using its existing theories of sets, differential, real and complex numbers. Proposed formalizations and corresponding case studies of smart grids are formally verified in the sound core of HOL Light theorem provers interactively.

The formalization of stability theory consists of a formal definition of stability in higher-order logic and also formal root analysis of characteristic equations, upto the fourth order, that are used for representing the control systems in the complex domain. This logical framework is then employed to formally specify and verify the designs of an H^{∞} current, H^{∞} voltage and H^{∞} repetitive current controllers for the power converters of wind turbines in smart grids.

© The Author(s), under exclusive license to Springer Nature Switzerland AG 2022
A. Ahmed et al., *Formal Analysis of Future Energy Systems Using Interactive Theorem Proving*, SpringerBriefs in Applied Sciences and Technology,
https://doi.org/10.1007/978-3-030-78409-6_6

To formalize the microeconomics models, i.e., cost and utility functions, we formally model the strict convexity or concavity and first-order condition in higher-order logic. To analyze the properties of cost and utility modeling functions, we also formally verify the second-order derivative test for strict convexity or concavity in higher-order logic. These formalizations, along with the formal root analysis, differential and convex theories of HOL Light, are employed to formally verify the properties of the polynomial type of cost and utility models upto the fourth order. This generic framework is then utilized to formally verify the electricity dispatch problem, in electricity market and quadratic utility model of DR programs.

Finally, we developed a logical formalization of asymptotic notations and formally verified their properties to assist the formal analysis of smart grid algorithms. To illustrate the usefulness of the proposed approach, we utilize it to formally analyze the computational complexity of two state-of-the-art PEV scheduling algorithms, namely, the Online cooRdinated CHARging Decision (ORCHARD) and online Expected Load Flattening (ELF) algorithms.

Mechanized proofs of stability, microeconomics and algorithm design theories, in this book, can be regarded as accurate due to exact modeling of corresponding mathematical models in higher-order logic. Moreover, due to an exhaustive analysis within the sound core of HOL Light theorem prover, no corner cases are left undetected and thus resulting in the explicit specifications for validity of given system models. Another major advantage of theorem proving is that mechanized proofs are easily reusable and tractable for the future use.

6.2 Future Work

The formalization, presented in this book, facilitates theorem proving based precise analysis of smart grid problems that are mathematically modeled using stability, microeconomics and asymptotic notations. Due to generic nature, these formalizations can be, directly or with suitable extensions, employed to many other smart grid problems and systems based on these mathematical notions. Some possible future directions are outlined below:

- The notions of marginally stable and unstable root analysis also play a vital role in the control system design and analysis. In this regard, the stability formalization, in Chap. 3, has already provided a formal reasoning framework which can be used to formally verify the conditions of marginal and unstable roots for characteristic equations upto the fourth order.
- In microeconomics theory, different mathematical functions are employed to model the behaviors of the economy agents such as logarithmic and exponential. The formal behavioral modeling, presented in Chap. 4, containing the formalization for the cost, utility and profit-maximization analysis provides sufficient formal support to extend the formal behavioral modeling using many other interesting mathematical functions.

- The formal asymptotic analysis, presented in Chap. 5, is used to formally verify the polynomial type of computational complexity of online PEV scheduling algorithms. However, proposed formalization can be used to formally verify the asymptotic bounds expressed as logarithmic and exponential functions, as well, which are commonly employed to design efficient algorithms for many safety or critical applications.
- Factorization of polynomials upto the fourth order, in Chap. 3, and strict convexity or concavity, in Chap. 5, are widely used in various engineering and science disciplines, and therefore, can be utilized to formally verify problems related to these fields. As most of the engineering systems can be represented using differential equations upto the fourth order, therefore, the proposed framework covers a wide spectrum of engineering applications. Moreover, the model reduction techniques can also be employed in case of the stability analysis of higher-order characteristic polynomials to reduce the order of the characteristic polynomials. On the other hand, convex optimization techniques heavily rely upon the convex and differential theories, and therefore, the proposed formalization, presented in Chap. 5, can easily be used to formally analyze various problems based on the convex optimization.
- Recent approaches of integrating formal methods with conventional computer-based tools such as MATLAB and Maxima, and using state-of-the-art machine learning techniques for auto-formalizations will require a large corporal of mechanized proofs. In this regard, the proposed formalizations are also important due to their wide usage in the analysis of the engineering systems.

Printed in the United States
by Baker & Taylor Publisher Services